Science, Technology and Innovation Studies

Series Editors
Leonid Gokhberg
Moscow, Russia

Dirk Meissner
Moscow, Russia

Science, technology and innovation policies (STI) are interrelated and connected. They are important as innovation drives economic development and societal welfare. The book series aims to contribute to improved understanding of these interrelations. Interdisciplinary in coverage, the series focuses on the links between STI, business, economy and society. The series offers theoretical and practical relevance through studying conceptual and empirical contributions. Relevant topics include STI and its economic and social impacts, STI policy design and implementation, entrepreneurship policies, foresight studies, emerging technologies and technology and innovation management. The series is addressed to professionals in research and teaching, consultancies and industry, governments and international organizations.

More information about this series at http://www.springer.com/series/13398

Leonid Gokhberg · Dirk Meissner ·
Alexander Sokolov

Editors

Deploying Foresight for Policy and Strategy Makers

Creating Opportunities Through
Public Policies and Corporate Strategies
in Science, Technology and Innovation

 Springer

Editors
Leonid Gokhberg
Dirk Meissner
Alexander Sokolov
Institute for Statistical Studies and Economics of Knowledge
National Research University
  Higher School of Economics
Moscow, Russia

Science, Technology and Innovation Studies
ISBN 978-3-319-25626-9   ISBN 978-3-319-25628-3   (eBook)
DOI 10.1007/978-3-319-25628-3

Library of Congress Control Number: 2016938000

Printed on acid-free paper

This Springer imprint is published by Springer Nature
The registered company is Springer International Publishing AG Switzerland

Contents

List of Authors

Cristiano Cagnin is advisor at the Center for Strategic Studies and Management Science, Technology and Innovation (CGEE). He holds a PhD and has previously worked at the EU Commission DG JRC-IPTS as a scientific officer. Dr. Cagnin is an industrial engineer who has been involved in research, international collaborative projects and consultancy in innovation, business strategy, environment management and cleaner production, and foresight. He is currently engaged in sustainability, RTDI as well as in bridging foresight and anticipation research and practice.

Jonathan Calof is professor of International Business and Strategy at the Telfer School of Management at the University of Ottawa, and co-director of the Telfers Business analytics-and Performance management area. He is a prolific author with over 150 publications and more than 1000 speeches, seminars and keynote addresses around the world on intelligence and foresight. In recognition of his contributions, Dr. Calof has been given several honours including Frost and Sullivan's life time achievement award in competitive intelligence; Fellow award from the Society of Competitive Intelligence Professionals; Honorary Professor at Yunnan Normal University in China; Appointment to the International Advisory Board for the Russian Foresight Committee of HSE; Honorary member of the Russian Society of Competitive Intelligence Professional; Board of advisors for the Centre en Intelligence Economique et Management Stratégique (CIE'MS/Center for Competitive Intelligence and Strategic Management in Morocco).

Jennifer Cassingena Harper has been engaged with the Malta Council for Science and Technology since 1989 in various capacities. Until 2011, she was the Director of Policy, Strategy, FP7 and International Cooperation with core responsibility for the National Research and Innovation Strategy and Foresight and links with the European Union. She currently retains a part-time consultancy role with the Council. Dr Harper is active at European and international level as advisor, reviewer, speaker and expert group member. She has been a member of a number of EU DG Research High Level Expert Groups as chair and rapporteur. Until 2011, she represented Malta in Competitiveness Council and the JRC Board of Governors. She is currently a member of the EU DG Research Expert Group on Strategic Foresight for R&I Policy in Horizon 2020' (SFRI). She has published a number of articles and papers internationally, Innovation Policies in Europe and the US (Ashgate, 2003) and co-editing The Handbook on Technology Foresight (Elgar 2009).

Moonjung Choi has 14 years' experience researching at Korea Institute of S&T Evaluation and Planning (KISTEP), affiliated with the MSIP (Ministry of Science, and ICT and Future Planning). For 11 of years, she participated in national S&T planning-related work that includes technology foresight, technology level evaluation, and technology assessment. Her original training was in Food Science and Technology. She received B.S. and M.S. from Yonsei University in Korea and earned Ph.D. at the Ohio State University in the USA.

Han-Lim Choi is a Research Fellow, Creative Economy Innovation Center, Korea Institute of S&T Evaluation and Planning (KISTEP). He received B.S., M.S. and Ph.D. from Korea Advanced Institute of Science and Technology (KAIST), specializing in Aerospace Engineering. He worked for Digital Appliance Laboratory (DA Lab.) of LG Electronics from 2005 to 2009 as a Chief Researcher and Team Leader. Since 2009, Han-Lim has been working for KISTEP as a professional in national planning including technology foresight, technology roadmapping and technology level evaluation. He conducted the 4th Korean Technology Foresight, the unique national technology foresight over the whole S&T fields and Technology Level Evaluation on National Key Technologies 2010. He also conducted Trilateral Technology Foresight among Korea, China and Japan on Renewable Energy and Technology Level Evaluation on National Key Technologies 2012.

Alexander Chulok is Deputy Director of International Research and Educational Foresight Centre at National Research University Higher School of Economics (HSE), Moscow. Dr. Chulok is responsible for coordination of research activities in the field of national and sectoral and corporate science and technology foresight. His scientific interests include theory, methodology, and practices of analysis of global challenges and grand responses in STI, priority setting, social and economic development, roadmapping and scenario building, foresight evaluation and implementation into policy-making process. Dr. Chulok has more than 50 scientific articles and hundreds of successful presentations and reports for different stakeholders.

José Cordeiro studied engineering at the Massachusetts Institute of Technology (MIT) in Cambridge, MA, economics at Georgetown University in Washington, DC, management at INSEAD in Fontainebleau, France, and science at Universidad Simón Bolívar (USB) in Caracas, Venezuela. He is director of the Venezuela Node of the Millennium Project, founding faculty and advisor at Singularity University, NASA Research Park in Silicon Valley, founder and president emeritus of the Venezuela Chapter of the World Future Society, and former director of the World Transhumanist Association and the Extropy Institute.

Erkan Erdil is professor and the director of Science and Technology Policy Research Center (METU-TEKPOL) at Middle East Technical University (METU) in Ankara. He holds a Ph.D. from University of Maastricht. Prof. Erdil is the board member of GLOBELICS (The Global Network for the Economics of Learning, Innovation and Competence Building Systems). He teaches introductory economics, microeconomics, statistics, econometrics, and economics of technology and work organization courses. His main areas of interest are economics of technology, labor economics, applied econometrics, economics of information and uncertainty. He has been author/co-author, and referee of articles in American Economic Review, Applied Economics, Applied Economics Letters, Agricultural Economics, METU Studies in Development and presented papers to various international conferences.

Ricardo Seidl da Fonseca is affiliated with National Research University Hogher School of Economics, Senior Researcher, Practitioner and International Adviser. He graduated in industrial engineering at the Federal University of Rio de Janeiro, Brazil, and post-graduated in industrial economics (Ph.D./Dr.-Ing.) at the Technology University of Munich, Germany. Dr. Seidl da Fonseca is a former Senior Industrial Development Officer, Unit Chief and Programme Manager at the United Nations Industrial Development Organization (UNIDO), Industrial Policy and Private Sector Development Branch. He is a member of International Advisory Boards of the Foresight Centre, National Research University Higher School of Economics, Moscow, and the National Foresight Programme, Bucharest. Before joining UNIDO, he worked in senior positions at the Brazilian National Council for Scientific and Technological Development (CNPq), the Federal Secretary of Industrial Technology (STI/MIC) and the Agency for Projects Financing (FINEP).

Hadi Tolga Göksidan has earned an Industrial Engineering degree from Gazi University in Ankara in 2001; and since 2002, he has been working as a research assistant in METU—TEKPOL Science and Technology Policies-Research Center. Additionally, he has worked as Technology Transfer Office Director in TOBB Economics and Technology University and Coordinator at the Middle East Technical University in Knowledge Transfer Office (METU-KTO). Formerly, he was the founder of the first technology transfer office (industrial TTO) rooted in in İvedik OIZ. Managing many regional and sectoral development projects, still, he is studying on many national and international research projects as a full time researcher, titled on regional planning, clustering, networking and foresight studies.

Leonid Gokhberg is First Vice-Rector of National Research University Higher School of Economics (HSE)—one of Russia's leading research universities—and Director of HSE Institute for Statistical Studies and Economics of Knowledge. He holds Ph.D. and Doctor of Sciences degrees and is Professor of Economics. Prof. Gokhberg authored over 400 papers published in Russia and internationally, e.g. several monographs and textbooks for universities. He has co-ordinated dozens of national and international projects, e.g. those sponsored by various national authorities, regional agencies and industrial companies, as well as by international organizations such as the OECD, European Commission, Eurostat, World Bank, UNESCO, UNIDO, IIASA, etc., in the areas of S&T and innovation indicators, analyses, foresight and policies. Prof. Gokhberg is a member of OECD and Eurostat expert groups on indicators for S&T, Information Society and education; the International Advisory Board of the Global Innovation Index (WIPO/INSEAD), and several other national and international high-level advisory groups on technology foresight and S&T and innovation policies. He is also Editor-in-Chief of Moscow-based scientific journal Foresight and STI Governance indexed in SCOPUS, and an editorial board member at Technovation and Foresight.

Oleg Karasev is Deputy Director of the International Research and Educational Foresight Centre at the National Research University Higher School of Economics (HSE), Moscow. He holds a PhD in Economics. Oleg has long lasting experience in macroeconomic forecasting, foresight, development of corporate innovation strategies and innovation policy. Oleg managed and participated in many foresight studies at national and regional level, projects on development of innovation strategies for leading Russian corporations, studies of future innovation markets.

Jonathan D. Linton is the Power Corporation Professor for the Management of Technological Enterprises at the University of Ottawa, Head of the Science and Technology Studies Laboratory of the National Research University Higher School of Economics (HSE), Moscow, and Editor-in-Chief of Technovation. He holds a Ph.D. in Management Science, Schulich School of Business, York University and is a registered professional engineer. Dr. Linton's research focuses on operational concerns associated with emerging technologies; entrepreneurship; science, technology, and innovation policy, and sustainable supply chains. He is on the editorial boards of Foresight, International Journal of Innovation and Technology Management, Journal of Engineering and Technology Management and Technological Forecasting and Social Change.

Dirk Meissner is Deputy Head of the Laboratory for Science and Technology Studies at National Research University Higher School of Economics (HSE), Moscow and Academic Director of the Master Program "Governance of Science, Technology and Innovation". Dr. Meissner has 20 years' experience in research and teaching technology and innovation management and policy. He has strong background in science, technology and innovation policy-making and industrial management with special focus on Foresight and roadmapping, science, technology and innovation policies, funding of research and priority setting. Prior to joining the HSE Dirk was responsible for technology and innovation policy at the presidential office of the Swiss Science and Technology Council. Previously he was management consultant for technology and innovation management with Arthur D. Little. Dirk represented Switzerland and currently the Russian Federation at the OECD Working Party on Technology and Innovation Policy.

Ian Miles is Head of the Laboratory of Economics of Innovation at National Research University Higher School of Economics (HSE), Moscow, and Professor of Technological Innovation and Social Change at Manchester Institute of Innovation Research (MIoIR), Manchester Business School. He previously worked at the Science Policy Research Unit, University of Sussex, which he left (as Senior Research Fellow) in 1990 to move to Manchester.

Prof. Miles received First Class Honours BSc in psychology from the University of Manchester in 1969, and received a higher Doctorate in Social Science from the same University in 2011. The latter was based on his studies of service innovation and Knowledge-Intensive Business Services. Other areas of research include foresight and futures studies, information technology innovation, and social indicators. His publications include 12 authored books, 12 edited books, over 100 journal articles, and over 200 reports and book chapters, with many other publications.

Anastassios Pouris is Director of the Institute for Technological Innovation of the University of Pretoria. He received his PhD from the University of Cape Town (South Africa) in policy related issues. Prof. Pouris is member of ASSAf; the Energy Efficiency Component of the 12I Industrial Tax Incentives Review Committee of CEF; the Research, Innovation Strategy Group of Higher Education South Africa; the International Society of Scientometrics and Infometrics; the editorial board of the journal Scientometrics and the Panel of Experts of IMD World Competitiveness Report. He has testified in a number of Parliamentary Committees and he acts as consultant for the National Advisory Council on Innovation, Department of Trade and Industry; Department of Science and Technology; World Intellectual Property Organisation etc. Prof. Pouris has published 76 articles in international refereed journals and a large number of major policy related reports.

Portia Raphasha is a Director in Innovation and Technology Policy at South African Department of Trade and Industry. She has completed her undergraduate degree in biochemistry and chemistry as well as an honours degree in biochemistry, both at the University of Johannesburg, and an MBA through the Management College of Southern Africa. Portia Raphasha is responsible for ensuring policy support in industrial innovation and technology development. Previously she was engaged with the Department of Science and Technology (DST) where she worked mainly on promoting bilateral engagements between South Africa and Americas & Asian countries. Prior to the DST, she worked at the University of Witwatersrand as a Research Officer and in the private sector.

Greg Richards is currently MBA Director and Director of the Centre for Business Analytics and Performance at the University of Ottawa's Telfer School of Management where he teaches courses on Corporate Performance Management, Analytics and Management Consulting. Dr. Richards holds MBA and PhD degree. His current research focuses on the application of performance management techniques in public and private sector organizations, the use of information technology in delivering performance information, the role performance information plays in stimulating organizational learning and innovation, and the issue of "cognitive load" related to decision-making processes.

Ozcan Saritas is Research Professor at the National Research University Higher School of Economics (HSE), Moscow; a Senior Research Fellow at the Manchester Institute of Innovation Research (MIoIR), Manchester Business School; and the editor of Foresight: the journal of future studies, strategic thinking and policy. Dr. Saritas' research focuses upon innovation research with particular emphasis on socio-economic and technological foresight. With a PhD from the "Foresight and Prospective Studies Programme" of the University of Manchester, he has introduced novel concepts like "Systemic Foresight Methodology (SFM)", which has been applied successfully to address long term issues and Grand Challenges involved in Sustainable Development, Renewable Energies, ICTs, Nanotechnologies, Higher Education, and Developmental studies.

John Edward (Jack) Smith is an Adjunct Professor of Technology and Strategy at the Telfer School of Management at the University of Ottawa. He is President of TFCI (technology foresight collaborative insights) Canada Inc. a private consultancy with clients in Canada and Europe, and a Director of the Proteus Institute of Canada, a not-for-profit research organization specializing in human security and entrepreneurship. Jack Smith is Chair of the Foresight Synergy Network (FSN) of Canada, a member of the International Advisory Board for the APEC Centre for Technology Foresight in Bangkok and a member of the Technical Committee for the European Commission's Future Technology Assessment Conference. He retired from the Canadian Federal Government in 2010, having served as Director of Science & Technology Foresight for the Office of the National Science Advisor, Government of Canada amongst other senior foresight positions.

Alexander Sokolov is Deputy Director of National Research University Higher School of Economics (HSE), Moscow, Institute for Statistical Studies and Economics of Knowledge and Director of HSE International foresight Centre. His main professional interests are related to foresight, S&T and innovation priorities, indicators and policies. Prof. Sokolov is a tenure professor at HSE, he teaches foresight for undergraduate and postgraduate students. He authored over 120 publications in Russia and internationally, managed many foresight projects, including: Russian S&T Foresight: 2030; Foresight for the Russian ICT sector (2012); Innovation Priorities for the Sector of Natural Resources (2008–2010); Russian S&T Delphi Study: 2025 (2007–2008); Russian Critical Technologies (2009) et al.

Prof. Sokolov is a member of several high-level working groups at the OECD and other international organizations and serves for advisory boards at several international conferences and journals.

Konstantin Vishnevskiy is Head of the Department for Private–Public Partnership in Innovation Sector at the National Research University Higher School of Economics (HSE), Moscow. He holds a Ph.D. from Moscow State University, Faculty of Economics. Dr. Vishnevskiy has long standing experience in the development of roadmaps, the elaboration of foresight methodology and corporate innovation development programme, the integration of foresight into government policy as well as financial and econometric modeling. He authors about 50 scientific publications on long-term planning and foresight, roadmapping, macroeconomic regulation and government policy, programme of innovation development for companies and gave more than 50 presentations at conferences and workshops on foresight, roadmapping and innovations.

Steven J. Walsh is the Regents' Professor and Director of the Technology Entrepreneurship Program at the University of New Mexico Anderson Schools of Management. He received his BE, MBA and a Doctorate of Philosophy in Management of Technology, Strategy and Entrepreneurship at RPI. Prof. Walsh is the founding President for the Micro and Nano Commercialization Education Foundation (MANCEF). He is an area editor for two journals and has provided special issue editor service seven times. He was named as one of 25 technology commercialization all-stars by the state of New Mexico in 2005, and in 2006 won the life time achievement award for commercialization of Micro and Nano firms by the Micro and Nano Commercialization Education Foundation.

List of Authors

Cristiano Cagnin Center for Strategic Studies and Management in Science, Technology and Innovation, Brasília, Brazil

Jonathan Calof University of Ottawa, Ottawa, ON, Canada

Moonjung Choi Korea Institute of S&T Evaluation and Planning, Seoul, South Korea

Han Lin Choi Korea Institute of S&T Evaluation and Planning, Seoul, South Korea

Alexander Chulok National Research University Higher School of Economics, Moscow, Russia

José Cordeiro Venezuela Node Department, The Millennium Project, Avenida Rómulo Gallegos, Caracas, Venezuela

Singularity University, Moffett Field, CA, USA

Venezuela Chapter, World Future Society, Caracas, Venezuela

Erkan Erdil Middle East Technical University, Ankara, Turkey

Ricardo Seidl da Fonseca National Research University Higher School of Economics, Moscow, Russia

Hadi Tolga Goeksidan Middle East Technical University, Ankara, Turkey

Leonid Gokhberg Institute for Statistical Studies and Economics of Knowledge, National Research University Higher School of Economics, Moscow, Russia

Jennifer Cassingena Harper Policy and Internationalisation Unit, Malta Council for Science and Technology, Malta

Oleg Karasev National Research University Higher School of Economics, Moscow, Russia
Lomonosov Moscow State University, Moscow, Russia

Jonathan D. Linton University of Ottawa, Ottawa, ON, Canada
Institute for Statistical Studies and Economics of Knowledge, National Research
University Higher School of Economics, Moscow, Russia

Dirk Meissner Institute for Statistical Studies and Economics of Knowledge,
National Research University Higher School of Economics, Moscow, Russia

Ian Miles Manchester Institute of Innovation Research, University of Manchester,
Manchester, UK
Institute for Statistical Studies and Economics of Knowledge, National Research
University Higher School of Economics, Moscow, Russia

Anastassios Pouris University of Pretoria, Pretoria, South Africa

Portia Raphasha University of Pretoria, Pretoria, South Africa

Gregory Richards Telfer School of Management, University of Ottawa, Ottawa,
ON, Canada

Ozcan Saritas National Research University Higher School of Economics,
Moscow, Russia

Jack Smith University of Ottawa, Ottawa, ON, Canada

Alexander Sokolov Institute for Statistical Studies and Economics of Knowledge,
National Research University Higher School of Economics, Moscow, Russia

Konstantin Vishnevskiy National Research University Higher School of Economics, Moscow, Russia

Steven T. Walsh University of New Mexico, Albuquerque, NM, USA

Foresight: Turning Challenges into Opportunities

1

Leonid Gokhberg, Dirk Meissner, and Alexander Sokolov

For many years, foresight has been used as an instrument for elaborating forward-looking strategies and policies, primarily in the science, technology and innovation (STI) domain (Johnston 2002; Keenan 2003; Keenan and Popper 2007). It has become a frequently used concept for preparing governments, businesses, research institutions, universities and non-for-profit organizations across the world to address potential future challenges.

Theoretical and methodological studies, as well as analyses of best practice cases, have enriched foresight tools and their applications across a wide spectrum of fields and areas. Extending the scope of foresight beyond its initial exclusive focus on STI (and especially on R&D), by looking at socio-economic and environmental trends and taking account of skills for STI, entrepreneurship, and other cushy topics, has provided an important feedback to the design of anticipatory STI policies (Sokolov and Chulok 2012). Academics and practitioners agree that although each foresight exercise is in many ways unique, there are several major 'mainstream' approaches which provide meaningful lessons to learn (Meissner et al. 2013). Thus, foresight used as an instrument for strategic STI planning in companies usually has a comparably short time horizon (with the exception of the largest companies in the energy, aerospace and other sectors with long-term innovation cycles) is allocated fewer resources and engages fewer stakeholders than that undertaken by public bodies. Foresight produced by government agencies to identify priority areas for STI either at the national level or in individual sectors tends to cover longer horizons and have a broader scope, involving more stakeholders. National STI foresight studies are the most complex in this respect (due to the increased coverage of sectors, technological areas and scientific disciplines) and require significantly more resources.

L. Gokhberg (✉) • D. Meissner • A. Sokolov
Institute for Statistical Studies and Economics of Knowledge, National Research University
Higher School of Economics, 20 Myasnitskaya Street, 101000 Moscow, Russia
e-mail: lgokhberg@hse.ru; dmeissner@hse.ru; sokolov@hse.ru

© Springer International Publishing Switzerland 2016   1
L. Gokhberg et al. (eds.), *Deploying Foresight for Policy and Strategy Makers*,
Science, Technology and Innovation Studies, DOI 10.1007/978-3-319-25628-3_1

Foresight is most often applied to identifying future applications and markets and their subsequent demands for particular technologies. Therefore, the challenges analyzed vis-à-vis technological trends (such as market pull vs. technology push) enable both businesses and researchers to identify the directions needed for forthcoming actions (van der Steen et al. 2011). Governments are provided with a better knowledge of the fields of basic and applied science which should be supported in the long-term.

The main ambition for applying foresight for countries' STI policy formulation and also for corporations' strategy development is to reflect on potential changes which might impact the nations and its businesses. Hence foresight is implemented to raise awareness about the potential short-, mid- and long-term developments expressed as challenges and opportunities (King and Thomas 2007; Martin 1995). Many individuals and businesses will seek to develop routines which help them **avoid** the challenges, predominantly for a short-term period of time and not always successfully, rather than **addressing** them (Sokolov 2009). Consequently, an approach evolves which focuses on threats to individuals, companies and societies instead of stressing incentives to develop initiatives that tackle the challenges and create new opportunities that may last for a long run. Though at an aggregate level, namely at a policy level and at company corporate level, the challenges including their expected impacts and threats are better understood. Nonetheless, this paradigm eventually generates and supports individual passivity, whereby, despite watching and monitoring the development of the challenges, and despite experiencing their growing impact, many actors still remain inactive.

The issue now is to integrate these challenges in the strategic orientation of national STI and of companies and to derive suitable measures to meet them and most important to implement such measures. In this regard it is important to remember that innovation stems from people's activities which in turn are driven by their ambitions and incentives to search for new solutions. The latter are reasonably different between people including: curiosity, a personal drive to do something new, and also a sense of what psychologists term "internal control". It is the attitude that one can shape the world and that a challenge is there to be solved and overcome. Consequently, STI policy and company strategies should take into account the ambitions of economic actors engaged into various links of an innovation chain, from the very early stages of interventions.

STI policy has focused on—in addition to support for R&D—infrastructures, regulations, and framework conditions of national innovation systems. Occasionally, public perceptions of STI have also been taken into account. Although skills issues are frequently discussed in STI debates, little attention has been paid to underlying personal attitudes and characteristics of individuals in the STI system. But knowledge of individuals' behavior and routines helps to achieve ambitious targets. This knowledge means that one can appreciate the provisional impact of possible STI policies. It is important nonetheless to understand not only people's motivations concerning STI, but also the potential objections and resistance towards proactive STI policy measures. The latter is especially important when it comes to policy actions which might affect established structures and routines referring to individuals, households, businesses, non-for-profit organizations, or governments (European Commission 2009).

Logically then, it follows that expectations towards STI actors continue to grow. The underlying assumption is that investment in STI generates economically viable innovation. Consequently STI policy aims to assign human, financial and material resources to selected fields of STI by setting respective priorities and by designing framework conditions that allow to enforce the exploitation and commercialization of science and public research. In light of the challenges detected and described by foresight and the desire to generate quick responses in particular, policy takes into account the fact that actual inspiration and academic freedom play limited roles. A switching of mindsets away from thinking in 'challenges' towards thinking in 'opportunities' therefore cannot be achieved merely by setting financial incentives, instead, this task requires to publicly recognize and reward individuals. In this respect, it is increasingly clear that STI policy and company strategies need to address the soft skills of human resources to design and implement initiatives addressing the challenges and that the private sectors credits the public sector overall contribution to enhancement of knowledge and science (Gokhberg and Meissner 2013; Meissner and Sokolov 2013).

Part I of the book discusses the potential and actual roles of foresight in the development of STI strategies, namely at a company level. The special features of national level **foresight are introduced in Part II. Part III highlights foresight in the broader STI policy context,** with a clear focus on switching the mindset from challenges to opportunities. The concluding **Part IV provides a framework for seizing opportunities for national STI development.** This final Part also provides an outlook on future developments of corporate and national foresight and how they could be implemented in innovation management and national STI policies respectively.[1]

Part I discusses **anticipatory strategies.** *Saritas* finds that monitoring trends is an important step for foresight activities and gives the first indications of emerging future developments in society, economy, and technology, and provides valuable inputs for future-oriented, strategic decisions at the levels of public and corporate policies. He considers how to integrate the results of the trend monitoring into processes of designing STI policy and business strategies. The chapter also spells out the practical aspects of how—and in what form—trend monitoring outcomes should be delivered to the target communities of policy makers and business planners.

The emergence of trends is naturally dependent on the diffusion of technologies and the role of stakeholders in the diffusion process. *Meissner* argues that STI strategies largely aim to support the diffusion of technologies and innovations in

---

[1] This volume complements an earlier book by the editors "Science, Technology and Innovation Policy for the Future: Potentials and Limits of Foresight Studies", Springer 2013. It summarizes the results of a high-level international conference "Foresight and STI Policy" hosted by the Institute for Statistical Studies and Economics of Knowledge, National Research University Higher School of Economics in Moscow, October 30–31, 2013.

commercially viable applications. The eventual impact of implementing these strategies is strongly influenced by a variety of stakeholders. However, the number and variety of stakeholders are not the only factors important for STI strategy; the agendas of stakeholders also matter. In the author's view, stakeholders may at first sight support the diffusion of technology yet their actual intent is different: the resulting activities potentially obstruct—instead of enforce—diffusion. The reasons for this are manifold. Frequently, while competing technologies are compared and competitive analysis is carried out, the overall infrastructure surrounding the technologies is insufficiently taken into account. This turns out to be a major barrier for technology diffusion and is driven by the stakeholders' hidden agendas. Hence in developing and implementing a technology diffusion strategy, it is important to systematically analyse stakeholders' agendas from all possible points of view.

The integration of foresight into corporate strategy-making raises special challenges. These include the compatibility of data and information collected through foresight with the standards required for corporate planning. The frequency of foresight and planning exercises is another issue. *Linton* and *Walsh* demonstrate how to integrate foresight with corporate planning as a way to help organizations understand what might be required in the future. Their chapter proposes a framework for determining the state of current and future competencies and capabilities of companies.

Setting the right priorities for STI activities is an issue of outstanding importance, especially for companies in knowledge intensive industries. The challenges mainly relate to how to build and maintain competencies for future oriented analyses of a company's external environment; how to achieve developments that have a positive impact on the companies' operations; and how to align the naturally different time horizons of corporate planning and future oriented strategic intelligence. The latter issue is particularly pertinent for commercial organizations in emerging and transition economies. *Vishnevskiy* and *Karasev* discuss the meaning of corporate foresight for innovation management and the interactions between corporate foresight and the corporate innovation process. They demonstrate the potential of corporate foresight for companies and also highlight the limitations of this approach.

Cordeiro provides an interesting comparison between the evolution of human beings and how this constant evolution causes ongoing changes in humans' routines. Changing routines, he argues, is mainly caused by evolution which uses technological progress as a tool for changing the status quo. A change—and hence technology—is not limited to narrow fields of application; it also causes secondary impacts which ultimately affect the broader set of routines. In this regard, we can assume that foresight and allied forward-looking activities potentially create a 'domino effect' on STI policy measures. In other words, it could be that foresight results have broader impacts on policy measures than usually expected.

Part II provides an insight into **different national foresight approaches in transition countries**. The chapter involves a rare collection of foresight studies undertaken at a national level. Governments in transition countries seem to be aware of the potential of foresight for designing national STI policies and for analyzing the strengths and weaknesses of national innovation systems. On the other hand, there is a widespread belief among stakeholders that their activities are sufficient for their country and that global trends have no (or only slight) impact on national innovation systems.

In their review of the process and results of foresight exercises aimed at identifying research priorities in South Africa, *Pouris* and *Raphasha* illustrate this contrast between government and stakeholder perceptions. They argue that national stakeholders in South Africa do not recognize the importance of emerging technologies and their respective impacts on economies and nations at large. foresight studies carried out in South Africa clearly show how the country is integrating itself into the global economy and is beginning to create awareness among key stakeholders about these developments and the need to identify national policies that respond to the resulting challenges.

Brazil has designed foresight in a way that explicitly positions societal actors as those able to develop the innovation system in directions that are crucial for addressing future challenges. *Cagnin* provides an insight into special Brazilian foresight features such as promoting transformative change to increase the relevance of foresight and its impact on decision-making processes and on the design and implementation of STI policies in Brazil. The Brazilian approach is intended to spark the imagination and expand collective understanding to better comprehend the present situation. It is assumed that this thorough understanding of the situation provides a solid platform for implementing policy measures to reorient the country's national innovation system. Achieving this ambitious goal requires a broad range of different competencies and positive attitudes of the actors involved to realistically assess the status quo.

To bring the relevant competencies together for a comprehensive assessment of the current situation and the potential development paths, the Russian Federation has developed and implemented a National Technology Foresight System. This is the subject of the chapter by *Chulok* who shows that a national foresight system integrates numerous actors with different affiliations from the country's existing competence centres. These are methodologically supported and coordinated but not centrally managed, and thereby decentralized competencies are leveraged. In addition this encourages competition between those specialized centres; which in turn also provides leverage for quality assurance of the respective foresight activities. The challenge imposed by such national systems of combined expertise is to ensure that the independent units follow similar approaches of foresight and that the results are comparable. Moreover, a national inventory / depository of foresight studies carried out by decentralized units would be beneficial and make the knowledge and experience acquired by these studies publicly shared and accessible to a broad national network.

In recent decades, South Korea emerged into a high tech country with a reasonable number of global industry leaders in several technology and innovation fields. This achievement is traceable—at least in part—to the remarkable history of foresight at the national level which was used for STI priority setting in all relevant spheres. *Moonjung Choi* and *Han-Lim Choi* explore how foresight in the entire field of science and technology has become a key process in national STI policy, resulting in key national initiatives such as the Science and Technology Basic Plan. The latter is not just a formal legal document, but a mandatory planning process established every 5 years by the Korean government, and it is the top-level plan shaping STI-related policies in Korea. It selects the national strategic technological priorities through reflecting on future technologies identified by foresight studies. The most recent South Korean foresight not only has a technological dimension, but also takes into account the development of society, its changing needs and desires, and the resulting implications for technology acceptance and diffusion.

Building a strategy that is related to STI is always done under a significant uncertainty regarding the intended outcome; therefore, it is a process associated with a reasonable risk. *Calof* and *Smith* argue in their chapter that while—at the moment of developing an innovation strategy—there might be demand for the intended outcome, this might change over time. For example, the demand could have been met by competitors. One approach to limit such risks is to integrate foresight, technology intelligence and business analytics into the initial design of strategies and to continuously monitor the external environment. Initially, this integrated approach was designed for companies' innovation management. Yet, the authors show that the integrated intelligence process also has potential for targeted STI policy.

Foresight and STI policy share several features. In principle, STI policy is targeted towards the future development of nations and societies by designing anticipatory policy measures which prepare countries for meeting future challenges at different levels. In this respect, STI policy should take an active role rather than merely reacting to current challenges only. STI policy measures certainly impact countries' STI but these impacts are frequently hidden and occur over a long time horizon. Decisions about and investments in STI priorities are always made under uncertainty at company and national levels. While foresight or similar activities have been already embedded in corporate STI strategies and priority setting, there has been still a lot to be done at country level. To date, it has become common practice in developed and emerging countries to use foresight for different purposes but the integration of foresight into the STI policy context remains a weak point. Therefore, **Part III** explores **the integration of foresight into a broader STI policy context**.

Using the example of Horizon 2020, *Harper* explores the potential of foresight and forward-looking activities in a STI support programme. She argues that foresight takes numerous roles in the design and implementation of an impactful support programme. This is mainly due to the numerous iterations in the design process and the decomposition of one huge programme into numerous sub-actions, which are all case-specific and targeted to different challenges. To meet this challenge, foresight takes a strategic, instrumental and operational role in the design and inception of the STI policy measures. However, the design of foresight in light of the EU Horizon 2020 programme needs to be carried out in a way that is sensitive to respective national environments and specific framework conditions which apply there. Horizon 2020 is a significant STI policy instrument designed and implemented by a multinational institution which naturally also reflects the interests of member states to some extent. Although, the approach chosen is not immediately transferable to countries' national foresight exercises, there are numerous positive lessons to be learnt by national policy makers.

Seidl da Fonseca provides an inspiring insight into the design and the final assessment of foresight at national levels by proposing a model for foresight assessment and for comparative analyses of STI foresight's impact. Particular country cases demonstrate a variety in methodological approaches and implementation schemes applied to foresight studies around the globe.

Each industry sector has particular features which require a dedicated tactics for futures thinking and foresight respectively. In particular, the services industry covers a broad range of different activities, and moreover, as *Miles* describes, beyond some traditionally recognized purely service activities there are also those which accompany manufactured products. In the latter case, services are thought to generate an additional value to a conventional product and hence provide a competitive advantage to the supplier. Both forms of services are close to the customer which means there is an opportunity to obtain an immediate user feedback. Services are also typically designed for the user and take into account users' wishes and requirements. Accordingly, Miles aligns foresight and futures studies to the features of services and the characteristics of innovation in services.

The capabilities of countries to meet global challenges and to turn a 'challenge-based thinking' into 'opportunity-based thinking', however, are not achieved at the national level. Rather, these capabilities emerge regionally. The exclusively regional (or even the city) level is much closer to value creation than the rather abstract, national (or federal) level. In fact, local networks are essential ingredients to broader value chains which may even obtain a global dimension. *Erdil* and *Goeksidan* show the potential for small and medium-sized companies of participating in global markets by means of integrating in local value chains. Such value chains display the local or regional networks which frequently change in their shape and orientation, and which often determine the overarching national competitiveness. Accordingly, these networks frequently assess their competitive positions and, more importantly, look for indications of future trends which might offer them new options to participate in global market activities.

In the **fourth concluding Part** *Gokhberg and Meissner* look at ways to benefit from STI. They argue that although there remains a need for designing a consistent and coherent STI policy approach and policy mix, the real challenge is to change the perceptions of the functioning of STI which is a pre-condition to achieving social and economic impact and value. This change is a shift from 'Thinking in Problems' which is characteristic of scientific work towards 'Thinking in Opportunities'. The latter still describes forward-looking activities but comprises of decomposing problems, searching for dedicated solutions, and developing necessary interfaces for integrating the latter into systemic strategies which are applicable to the initial agenda and not targeted at features of separate problems being taken on an individual basis, as usually implied in the 'Problem Thinking' mentality. The issue of changing mentality needs to be addressed at a policy level as well as by the STI communities.

Acknowledgements The book and this chapter were prepared within the framework of the Basic Research Programme at the National Research University Higher School of Economics (HSE) and supported by a subsidy granted to the HSE by the Government of the Russian Federation for the implementation of the Global Competitiveness Programme.

References

European Commission (2009) Policy mixes for R&D in Europe. UNU-MERIT, Maastricht

Gokhberg L, Meissner D (2013) Innovation: superpowered invention. Nature 501:313–314. doi:10.1038/501313a

Johnston R (2002) The state and contribution of international foresight: new challenges. The role of Foresight in the selection of research policy priorities, 13-14.

Keenan M (2003) Identifying emerging generic technologies at the national level: the UK experience. J Forecast 22:129–160

Keenan M, Popper R (2007) RIF (Research Infrastructures Foresight): Practical guide for integrating foresight in research infrastructures policy formulation. European Commission, Brussels

King DA, Thomas SM (2007) Taking science out of the box—foresight recast. Science 316:1701–1702

Martin BR (1995) Foresight in science and technology. Technol Anal Strat Manag 72:139–168

Meissner D, Sokolov A (2013) Foresight and science, technology and innovation indicators. In: Gault F (ed) Handbook of innovation indicators and measurement. Edward Elgar, Northampton, Cheltenham, pp 381–402

Meissner D, Gokhberg L, Sokolov A (2013) The meaning of foresight in science technology and innovation policy. In: Meissner D, Gokhberg L, Sokolov A (eds) Science, technology and innovation policy for the future—potentials and limits of foresight studies. Springer Heidelberg, New York, Dordrecht, London, pp 1–7

Sokolov A (2009) Future of S&T: Delphi survey results. Foresight-Russia 3(3):40–58 (in Russian)

Sokolov A, Chulok A (2012) Russian Science and Technology Foresight—2030: Key Features and First Results. Foresight-Russia 6(1):12–25 (in Russian)

van der Steen M, van Twist M, van der Vlist M, Demkes R (2011) Integrating futures studies with organizational development: Design options for the scenario project 'RWS2020'. Futures 43:337–347

Part I

Foresight for Anticipatory STI Strategies

Integration of Trend Monitoring into STI Policy

2

Ozcan Saritas

2.1   Introduction

The recent decades of innovation studies have been largely devoted to seeking the instruments of how countries and corporations can boost their socio-economic and technological performance while recognizing the differences between various innovation systems. The identification of the close relationship between technological progress, economic growth and societal well-being has attracted increasing attention both from scientific and expert communities as well as political and business leaders around the world. After several technological waves (information and communication technologies, biotechnology, nanotechnology and others), it has been observed that some countries have significantly increased their economic and political power, while the others have been followers or left behind. Consequently, the importance of technological progress has been noted both by scholars and practitioners who witness a rapidly changing world where breakthrough products and services create new markets and cement the dominant position of the leaders.

In this context, it has been viewed that mid- and long-term planning as well as the study of long term futures play critical role in ensuring successful national and business development. Policy makers, business leaders, and experts engaged to elaborate tools and methods that would allow them to look into the long term future, develop long term scenarios, articulate visions and roadmaps for strategies. Drucker (1964, pp. 15–16) noted:

> That executives give neither sufficient time nor sufficient thought to the future is a universal complaint... The neglect of the future is only a symptom; the executive slights tomorrow because he cannot get ahead of today... The future is not going to be made tomorrow; it is

O. Saritas (✉)
Institute for Statistical Studies and Economics of Knowledge, National Research University Higher School of Economics, 20 Myasnitskaya Street, 101000 Moscow, Russia
e-mail: osaritas@hse.ru

© Springer International Publishing Switzerland 2016

L. Gokhberg et al. (eds.), *Deploying Foresight for Policy and Strategy Makers*, Science, Technology and Innovation Studies, DOI 10.1007/978-3-319-25628-3_2

being made today, and largely by the decisions and actions taken with respect to the tasks of today. Conversely, what is being done to bring about the future directly affects the present. The tasks overlap. They require one unified strategy.

Since the beginning of the 1960s, foresight has been increasingly understood as a systematic and participative policy and strategy making approach with a long term perspective beyond usual time horizons for planning. One of the main motivations is, as Drucker stated, to create the future and bridge the future with the present. Foresight practitioners typically look into next 5–50 years, or longer depending on the focus of the study, to search for signals of change, and to identify opportunities and threats which the future may bring. Future-oriented knowledge gained through foresight is used to generate ideas about innovations and emerging technologies, which will be demanded in the future. This intelligence will be applied by the countries and companies, which wish to be on the leading side of technological, social and economic developments. The value of foresight exercises also lies in the way that the activities are inclusive and participative and bring a wide variety of stakeholders involved in science, technology and innovation (STI) and R&D systems together for collective visioning and mutual learning towards and common vision and priorities. The strategic decisions are made through consensus in a transparent process.

The process consists of a set of consecutive actions, including: (1) Intelligence (scoping, surveying), (2) Imagination (creativity, modelling), (3) Integration (visioning, priority setting), (4) Interpretation (planning, strategy making), (5) Intervention (action) and (6) Impact (evaluation), with a continuous process of (7) Interaction (participation) (Saritas 2013).

The process begins with the Intelligence phase, which involves a comprehensive understanding, scoping and scanning exercise. This is the phase, which created foundation for an evidence-based inquiry in a foresight exercise. Trend monitoring activities lie right at the beginning of this process. Monitoring allows capturing the major trends, which may be observed in social, technology, economic, environmental, policy and values/culture (STEEPV) systems, and may have potential implications for STI policy and strategy development. Most promising STI areas can be identified through the monitoring activities and can be prioritized based on their potentials to boost growth in the next decades while generating wealth, improving quality of life and environment.

Therefore, it is important to understand how the results of global trend analysis can be integrated into the actual policy making and strategic planning process to ensure that nations and companies make the right decisions in their pursuit of global leadership. The present chapter investigates the mechanisms of such integration and proposes analytical frameworks that would point potential gates of policy system in which the findings and visions emerging through trend monitoring can be incorporated into the actual policy making and strategic planning process.

Trend monitoring is frequently associated with technologies. This is due to the assumption that technology is one of the key drivers of progress in many spheres of life and it has a strong role in the process of national and corporate development and analyzing the socio-technical interactions and national policy systems. Many technological trends provided by the trend monitoring activities have national importance given their scale and large impact on the transformation of entire national innovation systems (NISs), which makes it critical to study how the knowledge about these technological systems can be integrated into the national innovation policy as well as corporate strategies. It should also be noted that the framework can be extended to cover not only technology trends, but also broader socio-economic trends. Therefore, the term 'trend monitoring' is used to cover identification, description, assessment and communication of all trends, which may be of relevant for public or private policy purposes.

Thus, Sects. 2.2 and 2.3 of the chapter embed the current topic in the innovation literature. Section 2.2 argues why technology and innovation are becoming increasingly important in the process of societal development and economic growth both at the national and corporate levels. Section 2.3 looks deeper into the national and technological innovation system approaches and notes the complementarity of policy approaches that demonstrate the importance of focusing both on systemic and technology-specific policies that are largely guided by the results of foresight and understanding of which STI areas are worth investing today.

Section 2.4 suggests an analytical framework for integrating the results of trend monitoring into the S&T policy making process. It studies the policy system and estimates the role of trend monitoring as an exogenous factor.

Section 2.5 looks at the linkages between the results of trend monitoring and business strategy planning processes. It also suggests a systemic view where trend monitoring is exogenous and guides the business strategists to make the right choices and decisions.

Section 2.6 elaborates on the practical aspects of delivering the results of trend monitoring to the policy makers and business planners in terms of how and in what form the results should be communicated to them and can be further used in the policy and strategy planning process.

The final section concludes the chapter by summarizing the major arguments. The major findings emphasize the critical importance of trend monitoring for the national and corporate development because the knowledge gained through this exercise is tightly bound to the issues of economic and corporate growth and societal well-being. The trend monitoring is closely related to the needs of policy makers and business planners and has multiple impacts on the policy and corporate planning process. The chapter finally presents various ways of communicating the results of technology monitoring.

2.2   Theoretical Foundations

Saritas and Smith (2011) consider trends as representatives of the broad forces and complex factors, involving diverse actors that lead and cause change in STEEPV systems with dynamic characteristics, until succeeded by others. They are typically experienced by everyone or a majority in more or less the same contexts insofar as they create broad parameters for shifts in attitudes, policies and business focus over periods of several years, and may have a global reach.

Monitoring trends is essential for STI development, which has firmly entered into the policy agendas and business strategies of the world's top nations and companies in recent decades (Mikova and Sokolova 2014). Today, the majority of policy makers understand the importance of promoting the advancements in the STI domains on a permanent basis, and falling behind in the pace of innovation is viewed as a definite failure in the national or business development.

Several theoretical and empirical reasons can be suggested for positioning STI development in the heart of countries' economic growth and societal well-being. From the economic theory perspective, Solow (1956) and Romer (1986) discussed the significance of technical progress in economic growth. Solow concluded that only 10 % of growth in the United States in 1909–1949 was ascribed to the increase of capital per worker while the rest 90 % were due to a variety of factors with a prominent impact of technical progress. Romer (1986) further endogenized the factor of technological advancement in his model of economic growth, which made the role of macroeconomic, science, technology and innovation policy yet more important in promoting national development.

Much earlier Schumpeter (1942) named innovation as the major driving force of market economies. The capitalist growth is driven by technological leaders and a large group of imitators who push the countries to develop. The stages of economic growth are replaced one after another by the process of creative destruction when a monopoloid company beats its competitors by destroying the present market conditions and creating an absolutely new reality where it possesses indisputable competitive advantages.

Proceeding from these and other economic findings, Freeman (2002) and Lundvall (2007) picked up on these and worked continuously to integrate the concept of the NIS, discussed below, with the theory of economic growth. Freeman (2002) studied the British, US and catching up countries' NISs and concluded that "technical change and the institutions which promoted it played a central role both in the forging ahead process and in the catching up process" (p. 208).

The Russian scholars have also noted the great importance of technological change in promoting political and economic leadership. Bogaturov (2006) identified five key factors that define the world leadership of nations: (1) military force, (2) economic might, (3) organizational resources, (4) science and technological strength, and (5) creative potential. All these factors are seen as highly dependent on the national innovation capabilities and therefore provide a clear link between the innovation potential and national leadership in the contemporary world.

Military force of state and non-state actors depends on the level of military and dual-use technologies. Key international players continue to increase their military research and development (R&D) expenditures[1] (e.g. US—from $43.3 bn. in FY2000 to $71.9 bn. in FY2007). Persistent growth of investment in the sphere led to creation of new weapons, which hold incomparable advantages over former weapon generations.

Economy is also dependent on innovative potential of states and non-state actors. States dominated by secondary and service industries are far ahead in comparison with those where the primary sector prevails.

Organizational resource is a complex of soft-power instruments used by states and non-state actors to strengthen their position on the world stage. In this regard one group of actors may use its sophisticated ICTs to manipulate public opinion and wage effective information wars. In contrast, the rest of the world, which does not possess any advanced technologies, practically ends up out of the world political process as it proves to be unable to establish coalitions of states and unite specific groups of people.

STI strength is by definition dependent on innovation. Public and private investment in high-tech industries brings much dividend, and successful start-ups turn into large corporations with significant level of capitalization.

Finally, creative potential is an ability of state, transnational corporations (TNCs) or other non-state actors to generate innovative ideas, which have enough market value to become a profitable product, process or service. Possession of such potentials gives incommensurable advantages to global leaders in inventing advanced technologies and making great scientific discoveries.

In the private sector research, development and innovation have also attracted much attention on the part of business and management scholars. The S-curve studies have proven that only technological excellence can allow companies to lead the market (Harvard Business Essentials 2003). The model (Fig. 2.1) vividly shows that if companies are unable to predict the emergence of new products and services they are bound to fall behind.

The problem becomes even more topical given that many technologies and products require long time to develop. For example, the aviation industry has the product life-cycle of 15–25 years, while microelectronics changes every 7–9 months. Paradoxically, the shorter is the product life-cycle the harder it is to compete and keep up with the leaders—i.e. if a company lags behind for a year it has already missed at least two product cycles while in aviation it would have an opportunity to accelerate the development in the remaining 14 years.

Thus, socio-economic and technological progress and national development are closely interlinked. This explains the increasing attention of the national governments and companies around the world for monitoring trends in all domains

---

[1] US Department of Defense's expenditures for military research has reached 63bn USD in 2014. http://archive.defensenews.com/article/20140420/DEFREG02/304200006/DoD-Reshapes-R-D-Betting-Future-Technology. Accessed 9 August 2015.

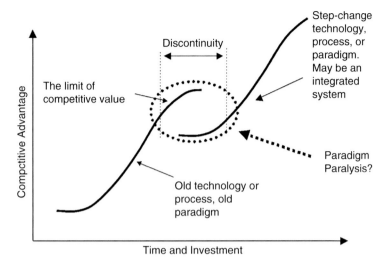

Fig. 2.1 S-curve: a business view of technological change. Source: Hinks et al. (2007)

of society, technology, economy, environment, politics and values/cultures and exploring opportunities and threats for STI development. Whether undertaken by public or private institutions, trend monitoring involves:

- Trends analysis with the examines a trend against its nature, causes, speed of development, and potential impacts.
- Trend analysis monitors specific trends with particular importance to a specific theme or sector and reports to key decision makers.
- Trend projection requires the use of quantitative data and extrapolation to portray the evolution of trends through time from the past into the future.
- Trend simulation is concerned with the modeling of a set of trends as a system with their interactions, which may help to develop future scenarios.
- Trend strategy aims at developing policies and strategies for increasing the benefits or mitigating the negative impacts caused by the trends.

The following sections will discuss and provide models on how these activities can be undertaken for the purposes of public STI policies and corporate R&D planning.

2.3 National and Technological Innovation Systems: Complementarity of Policy Approaches

Since the 1990s many governments and international organizations have adopted the concept of NIS that helps them identify the major strengths and weaknesses of their innovation strategies and formulate better science and technology policies.

The NIS approach was developed by Lundvall (1985, 1992), Freeman (1987, 1995), Nelson (1993) and Edquist (1997), and has since been embraced by many research schools and international organizations such as the Organization for Economic Co-operation and Development (OECD), the European Commission, and UNCTAD.

In basic terms, the approach combined several theoretical constructs that explained the foundations of the innovation process at the national level. First, it provided a route from systems of production toward systems of innovation by integrating Leontief's input-output analysis with innovation and entrepreneurship. Secondly, it developed a better understanding of the international trade specialization by combining it with the studies of the home-market economic systems. Thirdly, it explained the role of interactive learning in various national contexts. Finally, the concept reinforced the role of institutions in fostering innovation.

As a result, the innovation researchers have got a neat model of the national innovation development explained by both the structural elements of the national economy as well as network and institutional interactions between its agents. Traditionally three major actors (the triple helix) have been assigned to play the primary role of producing and promoting innovation: the government, private sector and universities. In this system the government supports basic science, carries out research and development and formulates and implements research and innovation policies; private sector produces innovation, invests in the applied research and ensures competitiveness of the country in the domestic and international markets; academia conducts broad range of research, provides training to future innovators via universities and consults other players on policy and strategy issues. However, more recently society has emerged as a strong player in innovation. Better informed and networked society has now got the power to influence public and corporate innovation and to innovate itself. Therefore, it is also crucial to monitor the developments not only in technological and economic spheres, but also the ones related to the socio-cultural evolution.

The technological innovation systems (TIS) approach came as a separate concept in the 1990s and 2000s (Carlsson and Stankiewicz 1991; Carlsson and Jacobsson 1997). Carlsson and Stankiewicz (1991) suggest that the technological systems should be analyzed both at the structural and functional levels with specific focus on institutions, networks and actors. They define technological system as a "network of agents interacting in a specific economic/industrial area under a particular institutional infrastructure and involved in the generation, diffusion and utilization of technology" (p. 94).

More recently, there has been a trend to use the TIS approach more actively in analyzing technological transition and radical change in the socio-technical environment. Markard and Truffer (2008) specified the concept of TIS by including only those groups of actors, networks and institutions that are "supportive to the innovation process, i.e. that share the goal of furthering at least some variant of the socio-technical configuration." Although this approach makes the model more precise analytically, it practically fails to capture the dynamics of the system, where the actors can change instantaneously their attitudes towards any specific variation of the new technology by modifying their strategies, switching to techno-logical alternatives and redirecting the money flows due to competition, changing market configuration or other reasons.

Dolata (2009) continues with a discussion of transformative capacity, adaptabil-ity and gradual transformation of a technology for achieving sectoral development. Generic technologies can be considered to be strategic to support TISs. For instance, nanotechnologies can be instrumental to achieve transformation in multi-ple other disciplines and commercial applications due to its multi-purpose charac-teristic. Such technologies may push the entire innovation system onto a new development trajectory given its exceptionally large scale and scope.

Finally, the study presented in this chapter builds upon the notion of innovation policy that has emerged in the last two decades (OECD 1997; Smits and Kuhlmann 2004; Metcalfe 2005). Broadly understood as STI policy, it includes a set of various measures that serve the ultimate goal of promoting innovation and supporting innovation actors to increase their learning and interaction with other agents of the system primarily universities and research labs. The development of the NIS approach led to the emergence of the concept of system failure that prescribes the policy makers to repair the functions and institutions of the entire national innovation system rather than be guided by the simplistic market failure approach (Smits et al. 2010).

The problems remain with distinguishing STI policies from all other policy domains, especially industrial and education policies. Flanagan et al. (2011) pro-vide a deep study of the so-called 'policy mix' that focuses on "the interactions and interdependencies between different policies as they affect the extent to which intended policy outcomes are achieved" (p. 702), and aims to bring together the entire variety of government measures that have an objective of fostering innovation development.

2.3.1 Policy Complementarities

Proximity of the NIS and TIS approaches has lately led to an understanding that the generic STI policies aimed at shaping and improving the NIS and the technology-specific policy instruments should be viewed as complementary. Indeed, as already mentioned above, it is recognized that some technologies have such a large transformative capacity that they can push entire national economies onto a new development trajectory.

Technology-specific policies are definitely guided by the vision of which STI areas will be most valued in future and will ensure the best return on investment. In this regard the forward-looking vision and technology anticipation provide precious information about the future technological developments and identifies the key directions where the nations and companies have any chance of surpassing the leaders and forging ahead.

Certainly, the trend monitoring activity represents only one of the elements of successful STI policy making and should be complemented by a comprehensive and constantly updated analysis of the NIS, its strengths and weaknesses as well as other foresight methods and tools that will provide a more detailed vision of how the future should be shaped in an inclusive and action-oriented way.

In this context trend monitoring provides a good vision of mega- or macro-trends and STEEPV areas that are worth investing. However, it does not look into the micro-level of specific applications that should be supported by nations and companies through individual projects. Therefore, it is a great platform for seeing the future on the whole and clustering the most promising technological areas but its results require further study and development that foresight centres can provide by using their additional expertise and competences.

2.4 Analytical Framework for the Integration of the Results of Policy Trend Monitoring into the Process of STI Formulation

This model proposed in this paper considers a systemic interaction between the results of technology monitoring and the STI policy making processes. A model is devised with the purpose of capturing most important aspects of the STI policy formulation and view the results of trend monitoring as a major input into the policy making process.

For the purpose of the present work the output of trend monitoring is considered as an exogenous variable that influences the entire policy system in general rather than each of its individual elements in particular. Further work can be done to endogenize the factor of trend monitoring into the STI policy making process at every stage. However, this will require a much more detailed analysis and deeper research. It is worth noting that a similar work took more than 30 years to endogenize the factor of technological progress into the model of economic growth as described in Sect. 2.2. The same process was witnessed in the policy studies when it took the same 30+ years to open up Easton's 'black box' (1957) and start looking into every policy element individually. So, it is obviously not a trivial task.

It should be noted that this work focuses only on the policy formulation mechanisms while the integration of the results of trend monitoring into the process of STI policy implementation and policy evaluation could require further elaboration and action. The analytical model discussed here certainly includes certain feedback loops that are linked to the policy evaluation process. Moreover, it is recognized that the knowledge of the future gained through the trend monitoring activity influences all actors who design and implement the policy as well as those who appear to be the primary objects of the public activity at every stage of the policy process. However, the scope of the present study does not cover an in depth analysis of this interaction to the extent of studying the behavior and impacts of every particular element and stakeholder.

All in all, the model presented in Fig. 2.2 involves a cyclical policy system that can be described in the following way: Input (trend monitoring) $\rightarrow$ Prioritization $\rightarrow$

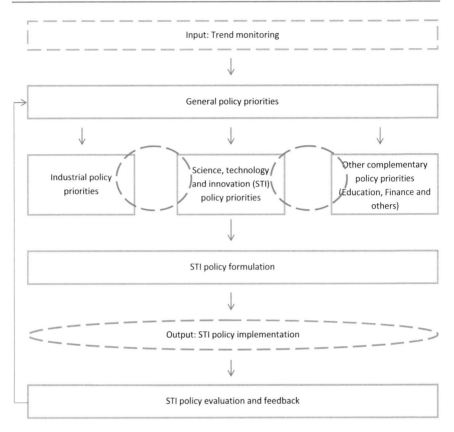

Fig. 2.2 An analytical framework of the integration of results of trend monitoring into the STI policy process. Source: adapted from Klochikhin and Saritas (2011)

Policy formulation → Output (policy implementation) → Policy evaluation → Feedback loop.

As already mentioned, the main focus is concentrated on the element of policy formulation while other components cannot be omitted from the analysis due to the complexity of political reality.

The study also recognizes the importance of strategic prioritization when governments define the major goals of national development and feed them into the actual policy formulation. Due to its macro nature, trend monitoring plays a critical role at every stage of the policy process providing the decision makers with knowledge about the future that can guide them in their choices and prioritization process.

Despite the purposefully simplified view of the model, it is obvious that the interaction between stakeholders at every stage of the policy process is not unidirectional. There are multiple feedbacks that are attributable not only to the entire system but are also observed at every stage of the process. For example, the general policy/national development priorities may be determined by political leaders. It is then the other relevant ministries, government agencies and other actors of the NIS take these into account those priorities as a guide when formulating priorities and policies in their respective domains.

In this context, future intelligence gained through trend monitoring is likely to influence the policy process in several ways. What is important to note that the process does not merely go top-down, but also bottom-up, where the long term visions and priorities of the top level are combined with the technical and practical knowledge, experience of the actors performing STI tasks, and the expectations of the broader society. Providing input for such an inclusive process, the monitoring activity considers a broad range of trends which may stem from or have impact on broader set of STEEPV systems.

Furthermore, similar feed-backs and feed-forwards between those who formulate policies and those who implement them will take place. The local practitioners, who start putting the policies into action interact with various agents (firms, research organizations, education institutions, etc.) that ideally participate in the process in the form of a foresight exercise, contribute for the shaping of the future STI policy and take active role in the implementation of results.

The actual impacts of the trend monitoring activities and the mechanisms of translating intelligence into action can be observed and measured at the STI policy evaluation stage. First of all, it is important to develop appropriate evaluation criteria for monitoring the impacts of trends. For instance, if the focus is on technology trends, indicators such as market size, technology readiness level, and R&D intensity can be used as parameters. However, there are challenges with the evaluation process too. Cause and effect relationships may not always be clearly identified, even if quantitative indicators such as market size are used. Therefore, the major difficulty for the evaluation phase is to give an objective and deliberate view of the policy implementation process. In such an effort, work undertaken by the evaluators might be unable to make an adequate assessment of what is communicated to them from different locations due to the influence of exogenous factors such as local cultural and institutional contexts that can vary from, for instance, one city/region to another.

Thus, there is much complexity involved in the policy process, and the results of trend monitoring might have enormous impact at every stage of policy making, implementation and evaluation. As the scope of the present study does not cover an analysis of the role of stakeholders in the system and the impacts of trend monitoring on each of them, it has the following assumptions:

- Every stakeholder operates with the same results of trend monitoring as all other stakeholders.
- No stakeholders has any additional knowledge about the technological area or economic sector where these results can be applied in the policy making process.
- The factor of time lag in communication of national priorities to the actual policy makers and further to the policy practitioners and other stakeholders is disregarded due to the fact that the rapid development of STI in the contemporary world can change the views of policy makers in a very short time period.

These assumptions allow the present study to present a generic framework, while writing off the issue of policy governance that can complicate it to a very large extent. Policy governance practically reflects the ability of policy originators to communicate their goals and aims to the stakeholders involved in implementing the respective public activity so that the latter embrace precisely the visions and policy tools prescribed by their leaders. These issues have been touched upon earlier during the discussion on the multiple feedbacks that guide the policy making process and determine the communication routes from the political leadership to the bureaucracy and vice versa. However, the area of policy governance presents so many challenges and uncertainties that it deserves a separate study rather than be discussed in full detail here.

2.4.1   Impacts

In basic terms, trend monitoring influences all stages of the policy process by providing intelligence about the future that is important in setting STI priorities, strategies and investments. More specifically, the following impacts of the trend monitoring activities can be mentioned on the policy making processes:

- Increasing awareness of existing and emerging trends: The social, technological or economic, monitoring trends gives opportunities to policy makers to stay up to date and ahead of developments; and to take the advantages or mitigate the impacts of what is likely to emerge in the future.
- Providing policy makers with tools for prioritising potential opportunities and threats and allocating resources to increase the ability to capitalise on, protect against, or mitigate the impacts of Grand Challenges and related potential disruptions. Thus, trend monitoring provides a solid foundation of awareness about trends, which may bring confidence to exploit benefits and take risks. Networking with key players by providing information about top countries, companies and institutions; funding organisations; potential collaborators and key people among the other stakeholders.
- Increasing the lead time to plan and address potential disruptions and making necessary political and strategic adjustments in the light of emerging trends and developments.
- Understanding trends to distinguish real trends from hypes.
- Ensuring higher level of stakeholder engagement in the policy making process by involving key stakeholders and wider society in trend monitoring activities. Consultations with practitioners, scientists, researchers, experts and other stakeholders give policy makers an opportunity to make more informed decisions.
- Making the policy making process more transparent. Trend monitoring provides objective information about the technological trends and usually makes it public. This information might prevent bias towards openly unreliable project

implemented by cronies and corrupt scientists while supporting genuinely scientific and prospective endeavors.

- Providing alternatives for technologies, strategies and policies. These may be unfamiliar to policy makers. The results of the project can point at technological substitutes when policy makers can choose one area out of three alternatives that would be most valuable economically and socially for a specific country.

2.4.2 The Structure and Functioning of Trend Monitoring for STI Policy Making

Aforementioned discussion aimed at showing the multiple impacts of the trend monitoring process almost at every stage of the policy process starting with the identification of general policy priorities at the top political level to more strategic and operational level decisions at the lower levels.

At the stage of general policy prioritization, trend monitoring provides valuable information to the political leadership, for instance, about the potential competitors in a certain S&T area, and prospects of a particular technology as compared with other alternatives. Given this information policy makers can stack up these prospects and results of competitive analysis against the social and technological capabilities of a particular country to understand whether they have enough potential to keep up with or even surpass the major competitors as well as make the best breakthrough in a particular technological area. At this top level the major outputs are usually the lists of critical technologies or priorities set forth by the political leadership.

The next level of prioritization goes down to industrial, STI, and education policies, most relevant to the issue of S&T progress. At this level, policy makers at the ministries and central/federal government agencies make a decision on the major priorities at the areas of their competence. Typically, these policy makers have much more special knowledge about particular sectors and technologies and can use the results of trend monitoring more efficiently.

The model presented here does not include any further disaggregation of the policy system to particular sectors (e.g. electronics, energy, chemistry, etc.) and regional policies. The impacts of trend monitoring on the policy making process at these levels are similar to the upper levels. However, it should be noted that the sectoral and regional policy making process has its own peculiarities, which are worth studying to increase the use of technology monitoring output and achieve higher impact.

In general, it would be proper to suggest that the level of ministries and other central government agencies should be the major target of experts running the trend monitoring activities. This proposition is explained by the focus of trend monitoring on the technologies of a relatively macro level without going deep into analyzing the particular applications of these technologies in certain products and services.

Meanwhile, focus here lies with the area of STI policy. As illustrated by the model in Fig. 2.2, STI policy represents one of the main areas of application of trend monitoring results together with industrial and education policies. All these policy domains represent a 'policy mix' aiming to promote innovation at the national level. In this mix policy makers usually interact between each other and set forth a range of overlapping priorities that serve the goals of creating an efficient national innovation system.

The STI policy formulation is usually carried out in the ministries of science and affiliated government agencies. The policy makers are aware of particular technologies and production methods that define the national innovation development. The major goal of these bureaucrats is to foster innovation proceeding from the assumptions put forward in Sect. 2.2, i.e. promote technological progress as an important means of supporting economic growth and societal well-being.

Technology monitoring activities provide further information about particular technologies and their alternatives that allow policy makers to understand what areas are more worthy investing rather than the others. Technology monitoring also engages broader expert community into the policy formulation process by putting the decision makers at one table with scholars and researchers who might have a say in defining the priorities and supporting particular projects. By this extensive engagement in the policy making process experts can make STI policy more responsive to the views of wider public and make the policy formulation more transparent with only most promising projects gaining needed support.

After the policy is formulated, it is communicated down the bureaucratic system where other professionals start implementing it engaging with many stakeholders whose interests are satisfied (or dissatisfied) by a particular policy. At this stage trend monitoring provides background knowledge about the technologies and their applications to the operational specialists who can use it to properly understand the policy details and requirements and make the proper actions while implementing the policy prescriptions.

In the meantime, the results of trend monitoring become critically important at the policy evaluation stage. The evaluators need to see the bigger picture of the technological development and estimate whether the decisions and choices were made correctly and communicated properly to the operational bureaucrats. Moreover, the policy evaluation would be more legitimate if it engages the expert community.

2.5   Integration of Trend Monitoring into Corporate R&D Strategy Planning

Although corporations and governments share many concerns when it comes to developing long term strategies and policies using broader foresight and trend monitoring activities, significant differences exist in terms of the scale of activities and how the future-oriented intelligence is used for strategy development and planning. In order to understand how trend monitoring may serve for R&D in corporations, it is important to first discuss the nature of innovation at this level.

The major difference lies in their primary missions: governments strive to shape an efficient national innovation system, while corporations usually aim at increasing their shares and profits in domestic and international markets. Sometimes the size of corporate operations may exceed the scale of a national innovation system, such as in the case of large multinational corporations (MNCs), such as General Electric, Siemens, or Unilever, which have a large variety of products and services in their portfolios.

In corporations, the nature, scale and culture of innovation are quite distinct compared to the national innovation systems. The visions and missions of corporations are better defined with a clearer scope in sectoral or service-oriented activities, whereas in national governments strive to manage all the sectors and services in the economy are focused in a broader stance with less clear boundaries. Higher precision of the commercial mission and corporate culture allows companies to sustain a better control over operations and employees, thus makes management of innovation relatively less complex and challenging than at the national level.

Corporations and governments also differ in their structural patterns. Nation states have to cover a large set of issues including social development, healthcare, and education. These areas of social life are extremely wide and concern a number of different actors with a wide variety of expectations. Companies typically have the opportunity to be more focused with a clear remit of promoting innovation and gaining the market leadership.

Considering the more precise and lean organizational structures of corporations and their R&D activities the following model can be suggested at this level (Fig. 2.3).

Due to the aforementioned differences between governments and corporations in terms of the scope and scale of innovation activities Fig. 2.3 illustrates a more focused and clear arrangement in the ways trend monitoring activities can be integrated in to the R&D planning processes.

The interactions between different components and functions in the system are more direct and dynamic, and this enables faster decision making and more efficient management of innovation activities. This is necessary for remaining competitive in fast evolving markets.

In corporations, trend monitoring activities may be undertaken by a dedicated strategy department with the participation of relevant corporate stakeholders ranging from different levels of management, operational units, suppliers, and clients among the others. The activities can be undertaken in conjunction with the corporate foresight activities. Ideally, the results are communicated equally to all corporate stakeholders irrespective of their position in the company, hierarchy and responsibilities.

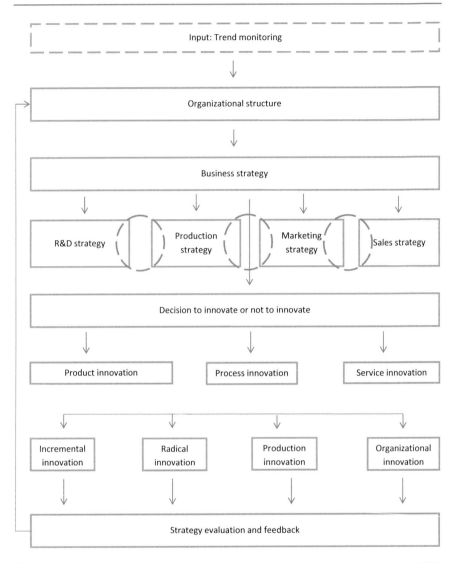

Fig. 2.3 An analytical framework for the integration of trend monitoring into corporate R&D planning process. Source: Adapted from Klochikhin and Saritas (2011)

2.5.1 Impacts

As in the case of the policy making process, the results of the trend monitoring create a normative environment that may give companies a number of opportunities. These may include:

- Generating ideas and identifying opportunities by studying trends by analyzing various sources of data (publications, patents etc.). Beating competition by anticipating emerging developments before the others and creating enough lead time for action.

- Balancing strategic goals by making decisions for longer term, which also helps to prevent any short-termism and low profit investments with no long term returns.
- Providing alternatives for technologies, strategies and policies. These may be unfamiliar to policy makers. The results of the project can point at technological substitutes when policy makers can choose one area out of three alternatives that would be most valuable economically and socially for a specific country.
- Assessing strategic options by benchmarking alternatives and possible market shares to be generated by them. The results of trend monitoring may hint corporations what choices they should make today to be the market leaders tomorrow.
- Developing new partnerships by studying other relevant products, services, actors and networks in the value chain. For instance, corporations might find potential partners in other countries and cluster around certain production method based on our analysis.

Providing stakeholder and society engagement in the process may prevent any serious backlash against the products and services provided by corporations.

2.5.2 The Structure and Functioning of Trend Monitoring for Corporate R&D Strategy

The corporate decision-making process includes multiple levels. The major decisions are typically made by a board of directors and concern the mission and organizational structure of a company. At this stage trend monitoring has strong potentials for influencing critical decisions. The corporate leaders might develop a strategy to create a new division, task force, or a project team to elaborate on a particular technology that is emerging and promising. Business leaders may consider re-allocating certain resources from one project to the other based on trend monitoring. Furthermore, new societal needs can be identified, as indicators of future markets, and some environmental concerns can be addressed as a part of Corporate Social Responsibility (CSR). Trend monitoring might create an environment that would homogenize the views across the company by providing an unbiased analysis of what technologies should be supported and which ones should be dropped. Therefore, the results might be used as an important argument in the corporate planning process. For instance, structural adjustments are usually tightly bound to the business strategy. It is a complex set of ideas and visions that guide the company to retain/gain market leadership. As a rule, business strategies comprise several sub-parts including the technology strategy, production strategy, business development strategy, marketing strategy, sales strategy, etc. All these strategies are represented and lobbied by particular divisions inside a corporation. They usually overlap in their understanding of the future and solutions to gain the best results but may also contradict each other significantly. It is a primary task of the chief executive officer to decide what opinions and ideas are most relevant to the corporate mission and trend monitoring can be a crucial support instrument for this purpose.

Another critical aspect of strategic planning is the decision on whether to innovate or not that is bound to the overall business strategy. This decision begins with a normative analysis provided by trend monitoring and helps to analyze if the company is already well-off with certain technologies or needs to invest more into the ongoing projects with a need for faster and more aggressive innovation in order to keep up with the competitors. If the company pursues a decision to innovate, then the type of innovation is to be decided—whether it will be a product, process, or service innovation, or any combinations of them. Trend monitoring will be able to help as a decision support tool to position the company, which strategy to pursue. This is an extremely critical decision, especially considering that innovation and R&D are usually rather time and resource-consuming. Making such critical decisions will require trend monitoring to be used with further competitive and market analysis conducted within the company and/or by other external consultants.

Like in the policy process, companies conduct regular strategy evaluations. These evaluations are crucial for companies given the highly volatile and dynamic market environment. Likewise the national level, at the corporate level, the results of trend monitoring can be fed into the evaluation process with some experts potentially employed as external reviewers.

2.6   Incorporating Trend Monitoring into Policy and Strategy Making Processes

Setting up a trend monitoring system is certainly an important first step to grasp existing and emerging developments in a wide variety of domains. However, it is not enough. Trend monitoring should be incorporated into the policy and strategy making process so that the results can be disseminated effectively by the clients of the activity. Typically, the trend monitoring team plays an instrumental role in transmitting the outputs through formal deliverables or by actively engaging in the policy and strategy making processes. The present study identifies four ways of incorporating the trend monitoring activity into existing structures. From more simpler and direct to more sophisticated and desirable levels, this incorporation may be in the forms of:

1. Communication
2. Participation
3. Experimentation
4. Integration

Each form of incorporation will be described below.

2.6.1 Communication

In this form of interaction the results of trend monitoring are provided to national and corporate policy makers through a direct communication. This can be done through open or confidential reports, publications, joint workshops and meetings, as well as by personal contacts and interactions. By means of this interaction trend monitoring results are publicized with the aim of providing information and intelligence about major trends to relevant parties, who can further use it in the policy making and strategic planning processes.

Trends can be communicated in the form of reports on trends or regular briefs about important and emerging trends. Information may be included about the title and description of trends; their impacts; where they originate from and where the impacts can be observed by considering the STEEPV systems; their relationship to other mega trends and Grand Challenges; geographical impact (i.e. global, national or regional); industries, sectors and research areas where the actions can be taken. Further information can be included to indicate the markets, products, technologies and R&D areas related to the trend; possible counter-trends; and enablers and barriers which may affect the trends' development trajectory, among all other relevant indicators depending on the needs of the national and corporate policy makers.

Disseminating intelligence about the existing and emerging trends will provide crucial input for future-oriented policy and strategies, competitiveness analysis, and identifying partners and designing networks for future collaboration. Companies can also use this knowledge to carry out a technology assessment and make better choices.

2.6.2 Participation

In this form of interaction the trend monitoring team is closely cooperating with policy makers and corporate planners. This may be in various forms. They may be working together as a team, or may play the role of consultants, reviewers, or policy/strategy evaluators. In this case the trend monitoring team has a more central role in policy and strategy making process. By this active participation into the policy and strategy making process, the technology monitoring provides input from a broader range of STEEPV systems and represent the visions and concerns of the wider public, stakeholders, and consumers in the case of corporations, who play a major role in the areas of concern. They deliver new knowledge and normative views to the policy makers in a more context aware and inclusive policy making process.

2.6.3   Experimentation

Trend monitoring may be also used at the later stages of the policy and strategy formulation to validate the findings and decisions. In this form of interaction members of the trend monitoring team and experts involved in experimental processes. They may participate as evaluators and reviewers before, during and after the policies and strategies are put into practice. They can also participate in pilot projects, living labs, or review the strategic choices made.

2.6.4   Integration

The role of integration is two-fold: First, the trend monitoring function is integrated physically into the policy making and corporate planning process, which goes beyond merely experimentation and evaluation. By integrating the trend monitoring function in the form of experimentation, a real time input can be provided through the feed-forward and feed-back of intelligence gained, which will allow more focused and target oriented monitoring efforts. Meanwhile, a real-time input can be provided to increase the evidence-base of policies and strategies. This is the type of interaction, which creates the highest level of impact from trend mining to policy and strategy making. The ultimate aim of trend monitoring activities should be to achieve this sort of engagement with a close proximity to decision making structures.

2.7   Conclusions

This chapter has developed a methodological framework for the integration of the results of trend monitoring activities into the processes of STI policy formulation and corporate R&D planning. Two models were provided that view the results of trend monitoring as an exogenous factor influencing all relevant stakeholders equally and disregarding the time lags in communication of the results to different levels of decision making. It is assumed that the model can be developed further to endogenize the factor of trend monitoring and study its impact on every stage of the policy making and business planning processes.

Overall, the impacts of trend monitoring on the policy making and corporate planning process include generation of intelligence about the future and creation of a normative environment that can guide government and corporate policy makers to formulate better decisions and priorities by considering what is likely to emerge; increase public and stakeholder engagement; provide better transparency of the policy processes and undertake more comprehensive competitive analysis.

In practice, the results of trend monitoring need to be delivered to the target audiences of policy makers and corporate planners. The present work has identified four different levels for this process, ranging from mere communication, to more active participation, collective experimentation, and efficient integration.

Depending on the purpose of the activity, the roles and interests of the stakeholder groups, and level of engagement needed one or more of means of communication can be considered for dissemination and feedback.

As a final point, it is important to highlight that trends data should be based on reliable sources. Although a wide variety of information sources ranging from structured scientific and academic databases to semi- or un-structured information in websites, blogs, social networks, or reports can be used during the trend monitoring activity, and a wide variety of tools and applications are available for bibliometric, scientometric and semantic tools, it is important to ensure that the source of information is credible and acclaimed to provide a better evidence base and less, or, if possible no, hypes by mainstream media and discussions. Due to increasing amount of information and data are produced, and the importance of anticipating what might emerge in the future is becoming more crucial, trend monitoring activities are becoming not an option, but a must for both public and private policy and strategy makers.

Acknowledgements The author's contribution was produced within the framework of the Basic Research Programme at the National Research University Higher School of Economics and was supported within the framework of the subsidy granted to the HSE by the Government of the Russian Federation for the implementation of the Global Competitiveness Programme. The publication draws upon an earlier work by Klochikhin and Saritas (2011), "Development of an approach for the integration of the results of the Global Technology Trend Monitoring into the S&T policy formulation process", an unpublished report produced for National Research University Higher School of Economics, Moscow.

References

Bogaturov AD (2006) Leadership and decentralization in the international system (in Russian). Int Trends 12(3)

Carlsson B, Jacobsson S (1997) Diversity creation and technological systems: a technology policy perspective. In: Edquist C (ed) Systems of innovation: technologies, institutions and organisations. Pinter, London

Carlsson B, Stankiewicz R (1991) On the nature, function and composition of technological systems. J Evol Econ 1:93–118

Dolata U (2009) Technological innovations and sectoral change. Transformative capacity, adaptability, patterns of change: an analytical framework. Res Pol 38(6):1066–1076

Drucker PF (1964) Managing for results: economic tasks and risk-taking decisions. HarperCollins, New York

Easton D (1957) An approach to the analysis of political systems. World Politics 9(3):383–400

Edquist C (ed) (1997) Systems of innovation: technologies, institutions and organizations. Pinter, London

Flanagan K, Uyarra E, Laranja M (2011) Reconceptualising the 'policy mix' for innovation. Res Pol 40(5):702–713

Freeman C (1987) Technology policy and economic performance: lessons from Japan. Pinter, London

Freeman C (1995) The national innovation systems in historical perspective. Cambridge J Econ 19 (1):5–24

Freeman C (2002) Continental, national and sub-national innovation systems—complementarity and economic growth. Res Pol 31(2):191–211

Hinks J, Alexander M, Dunlop G (2007) Translating military experiences of managing innovation and innovativeness into FM. J Facilities Manag 5(4):226–242

Klochikhin EA, Saritas O (2011) Development of an approach for the integration of the results of the Global Technology Trend Monitoring into the S&T policy formulation process. An unpublished report produced for the Institute for Statistical Studies and Economics of Knowledge (ISSEK), National Research University Higher School of Economics, Moscow

Lundvall BA (1985) Product innovation and user-producer interaction. Aalborg University Press, Aalborg

Lundvall BA (ed) (1992) National systems of innovation: towards a theory of innovation and interactive learning. Pinter, London

Lundvall BA (2007) National innovation systems—analytical concept and development tool. Ind Innov 14(1):95–119

Markard J, Truffer B (2008) Technological innovation systems and the multi-level perspective: towards an integrated framework. Res Pol 37(4):596–615

Metcalfe SJ (2005) Systems failure and the case for innovation policy. In: Llerena P, Matt M, Avadikyan A (eds) Innovation policy in a knowledge-based economy: theory and practice. Springer, Heidelberg, New York, Dordrecht, London, pp 47–74

Mikova N, Sokolova A (2014) Global technology trends monitoring: theoretical frameworks and best practices. Foresight-Russia 8(4):64–83

Nelson RR (ed) (1993) National innovation systems: a comparative analysis. Oxford University Press, Oxford

Organization for Economic Co-operation and Development (1997) National innovation systems. OECD Publishing, Paris

Romer PM (1986) Increasing returns and long-run growth. J Political Econ 94(5):1002–1037

Saritas O (2013) Systemic foresight methodology. In: Meissner D, Gokhberg L, Sokolov A (eds) Science, technology and innovation policy for the future potentials and limits of foresight studies. Springer, Heidelberg, New York, Dordrecht, London, pp 83–117

Saritas O, Smith J (2011) The big picture—trends, drivers, wild cards, discontinuities and weak signals. Futures 43(3):292–312

Schumpeter JA (1942) Capitalism, socialism and democracy. Harper& Row, New York

Smits RE, Kuhlmann S (2004) The rise of systemic instruments in innovation policy. Int J Foresight Innov Pol 1(1–2):4–32

Smits RE, Kuhlmann S, Shapira P (eds) (2010) The theory and practice of innovation policy: an international research handbook. Edward Elgar, Northamptonand Cheltenham

Solow RM (1956) A contribution to the theory of economic growth. Q J Econ 70(1):65–94

Identification of Stakeholders' Hidden Agendas for Technology Diffusion

3

Dirk Meissner

3.1 Technology Diffusion

Technology diffusion has long been discussed in academic literature. The main focus of the discussion and concept was aimed at the management of technology and its subsequent diffusion in the market, respectively in applications. More recently the discussion was extended to diffusion of technology from the science, technology and innovation (STI) policy perspective. Namely during the last decade policy makers have especially become more and more aware of the challenges of diffusing innovation in application on a broader scale which leads to the changing attitudes of politicians from considering only the development and generation of technologies and innovation to a more systemic view that includes generation and adaption, e.g. diffusion. At first sight this does not incorporate much change. However for STI policy and policy makers this imposes the challenge of designing and implementing measures that include the diffusion phase.

Although radical changes imposed by new technologies are often the consequence of new entries into competition, external shocks or crises, the outstanding performance of new technologies, market changes and/or industry competition (van den Hoed 2007), there still is no standard form of technology diffusion (Rao and Kishore 2010). In addition, there is a constraint for policy intervention at this stage in the technology and innovation life cycle: in liberal market economies, direct public intervention on the market is not allowed, e.g., the direct support of technology and innovation diffusion is only reasonable in the case of market failures. The difficulty however lies in the definition of a market failure.

In a broader sense, innovations are potential replacements of already existing solutions. Thus it is a matter of competitiveness of the innovation and its ability to

D. Meissner (✉)
Institute for Statistical Studies and Economics of Knowledge, National Research University Higher School of Economics, 20 Myasnitskaya Street, 101000 Moscow, Russia
e-mail: dmeissner@hse.ru

© Springer International Publishing Switzerland 2016
L. Gokhberg et al. (eds.), *Deploying Foresight for Policy and Strategy Makers*,
Science, Technology and Innovation Studies, DOI 10.1007/978-3-319-25628-3_3

outperform existing solutions and replace them, hence there already is an existing market. With increasing numbers of innovation users, more people, e.g. customers, become aware of the solution and start forming opinions based on the experiences of others (Zapata and Nieuwenhui 2010). In principle this is a reasonable assumption, but it still neglects the fact that competition is not necessarily limited to the actual technology or innovation, but instead replacement competition includes the broader environment in which a new technology can be used and operate (Meissner et al. 2013). Therefore technology assessments in the traditional way need to be extended to include the respective opportunity cost for investments that have already been made, so the assessment mostly centred on the technology's characteristics is complemented by an even stronger economic dimension.

At the policy level changes are initiated in technology regimes and they usually come with stakeholder involvement in the early phases of a new technology's diffusion into application. It is common knowledge that this requires the substantial involvement of them but there are also potential threats affiliated with this. Brown (2003) and Reed et al. (2009) argue that including stakeholders requires policymakers' skills in bargaining, negotiating and cooperating with those. Also Brown (2003) finds that such early involvement supports the building of absorptive capacities by stakeholders. Furthermore it also provides a platform to reach a consensus not only between policymakers and stakeholders, but also among stakeholders by exchanging different understandings and eventually validating a common understanding (Brown 2003). Accordingly policy needs to find alternative routes to influence the market and direct it towards the replacement of existing solutions (Meissner 2014).

Geroski (2000) finds that information and communication channels are suitable instruments for policy makers to influence the markets by involving stakeholders in the initial political agenda. Typically the stakeholders participation in such politically initiated and driven communication and information-related undertakings is accompanied by direct measures such as subsidies, which might emerge (Geroski 2000). Policy makers have a range of opportunities to influence the diffusion of technologies, namely the information about a technology and subsidies in different forms (Geroski 2000).

The communicative role of policy makers includes targeted and dedicated information for a broader audience, the willingness to perform and act as lead users as well as facilitating the communication between different stakeholders who are affiliated with the technology in question. Such communication activities often aim at building consensus among them about the characteristics and features of technologies and also at informing the potential user community about the technologies, thus trying to influence the perception of technologies and the attitudes towards them. This is mainly the case for disruptive innovation, which is typically a distinctive feature of premium products in markets that are often saturated with a reasonable share of mass produced products. Moreover especially capital-intensive industries are confronted with additional pressure on companies to assure timely return on of investment. Accordingly the uncertainty of the eventual acceptance and return on the innovation increases with a higher degree of disruptiveness of the innovation. Disruptive innovations that are included in an existing product have even more impact than the sole replacement of technologies/ solutions, because they require the revision and adjustment of the manufacturing and assembly process as well as the respective changes and adaptations in the supply chain. Thus besides the initial innovation, which can come in different shapes, additional innovations, often process innovations, are required, which are shown to be essential for eventual acceptance and diffusion.

Public support is frequently granted in different forms, be it either during the actual development and testing of technologies or at the stage of market penetration, e.g., application diffusion (Proskuryakova et al. 2015a, b). In principle subsidies are intended to set incentives for users (user subsidies), to accelerate technology development (push subsidies) and to stimulate competition between suppliers (competition subsidies) as well as to support interface harmonization and the standardization of technologies (technical norms subsidies).

The change of a complex system of actors requires changes of multiple elements of the system (van Bree et al. 2010; Proskuryakova et al. 2014; Schibany and Reiner 2014; Simachev et al. 2014). Basically these changes naturally open opportunities for all actors participating in the system, however the challenge remains to overcome the different perceptions of the opportunities that the actors have. These perceptions are individual, thus the actors in the system will define their strategies and act according to their own interest, which does not necessarily reflect other actors' strategic intentions. To overcome conflicting interests and enable technology and innovation diffusion, political skills such as bargaining, negotiating and collaborating can be helpful in aligning the stakeholders' interests (Brown 2003). Hence a profound understanding of the stakeholders' interests becomes an essential precondition. Over the last decade, stakeholder analysis has evolved as a broadly recognized and applied instrument for finding consensus between stakeholders, i.e. the elements of a system (Friedman and Miles 2006; Jepsen and Eskerod 2009).

Another typical characteristic of disruptive technologies is their applicability to a limited and narrow range of pioneering customers whose experiences with the initial application are commonly valuable and useful for further development, which in some cases leads to a broad mass application eventually. The broad introduction of disruptive innovations frequently requires a transition of the established industry, including significant changes in the socio-technical regime of the industry (Collantes 2007). Van Bree et al. (2010) propose a four element model for analyzing the technology diffusion using the example of automobile industry. First, there is a transformation stage during which niche technologies are in principle operational but still require additional development for mass application. Second, the industry is confronted with a sudden change in the industry environment which forces the industry actors to seriously consider changes in the existing technology regimes. Considering changes does not imply automatic replacement of existing technologies rather this means intensive competition between emerging niche technologies up to the point until one technology becomes dominant. Third, the substitution of technology takes place, and, fourth, the established set up of the industry is shaken and some actors are replaced.

The replacement of actors might include the replacement of original equipment manufacturers (OEMs) as well as significant changes in the supplier landscape. But also the scale of use needs to be seen in the context of the application, e.g. this does not automatically imply mass manufacturing for consumer devices but also relates to unique and dedicated applications for only a limited number of potential customers (Zapata and Nieuwenhui 2010; Kutsenko and Meissner 2013; Vishnevskiy et al. 2015). Also, predicting the potential diffusion of technologies requires an analysis of the whole picture, which naturally includes the technology in question but also the impact of the technology on other subsystems and on the actors involved in the actual technology and subsystems (Keles et al. 2008). Subsystems are understood as the systems that surround a technology and are essential for the latter operation and use.

3.2 Disruptive Innovation: Fuel Cells in the Automotive Industry

The STI policy perspective is especially important in light of the Grand Challenges to which clean energy belongs. During the last decade governments have initiated and supported substantial efforts to redesign the current national energy policies and priorities, switching the focus to renewable energies in several application fields. Increasingly these efforts are questioned and challenged by numerous stakeholders for different reasons.

The main obvious and most frequently cited motivation for raising concerns about green energy is the issue of cost, which is used to question the impact and effect of related research efforts. It can be observed that in many countries policy makers are confronted with the quest to justify the substantial investment already made, but even more they are confronted with comparisons of the investment required to change the energy mix favouring green energy over traditional fossil sources and the respective returns from this change. Often policy makers are stuck in a dilemma, which is to enforce the change in the energy mix by various instruments, but also limiting the burden on the energy consumer which in itself is challenging. For changing the energy mix, renewable energies are thought by politicians as being advantageous in many shapes. It is assumed that the technologies are already readily available or at least are available for application in the near future. Though the question about whether or not these new technologies will be accepted and applied in the market is not answered, e.g. the current or near term technology availability itself is no guarantee that technologies will diffuse in the market application.

The automobile industry is considered one application field for mobile fuel cells with significant market potential. It has become a common practice in the industry to introduce breakthrough innovations to premium products first, thus making them accessible to a comparably small share of customers in the early stages of the innovation life cycle (Zapata and Nieuwenhui 2010). Fuel cells have been in discussion as a substitute for combustion engines for a long time. However, the discussion was mostly around the replacement of internal combustion engines with different, alternative engines including fuel cells and most frequently hydrogen powered engines. Accordingly there were a reasonable number of assessments of fuel cells for different applications, including mobile and stationary applications (Hart 2000; Pehnt 2003; Baretto et al. 2003; Afgan et al. 1998; Afgan and Carvalho 2004; Hopwood et al. 2005; Midilli et al. 2005a, b). It has been recognized that energy sources such as hydrogen are required that do not cause societal impacts (Afgan et al. 1998; McGowan 1990; Hui 1997; Dincer and Rosen 1998, 2005, Hammond 2004). Schwoon (2008) finds that emissions from internal combustion engines have decreased substantially per engine recently but these reductions have been offset by the increase in car travel and heavy and light duty vehicles.

Radical changes such as fuel cells in the automobile industry are taking a long time to diffuse for several reasons (Collantes 2007). In the special case of fuel cells, it is evident that the technology per se has been available for quite a long time, but that there was no incentive for the industry to replace existing solutions, e.g. technologies. This is at least partially due to the modest demand for the technology which appears all the more surprising because the technology itself has been under discussion in the public domain for a long time and efforts have been made to establish the technology several times. Still, expectations towards the technology have been very high and could not be met, either in technological terms or in terms of performance of the final product used by the customers (Collantes 2007). Moreover customers have been used to vehicles powered by internal combustion engines for decades, which has resulted in routines that the customers developed for operating them. Changing routines is widely accepted as one of the most challenging and difficult undertakings of technological diffusion.

The diffusion of fuel cell vehicles is dependent on a variety of factors beyond the actual technology, e.g. fuel cell, and the respective product (fuel cell vehicle) which influence each other eventually forming a system around the actual technology (Rodriguez and Paredes 2015). Among these factors is the fuel cell operation which requires a manufacturing and maintenance infrastructure but also the fuel supply infrastructure which differs significantly from the existing fuel supply infrastructure (Keles et al. 2008; Turto 2006). The fuel supply infrastructure imposes the development and/or adaptation of norms and standards that take account of the special characteristics of the fuel itself, and its production, transport and storage. The challenge is that the different elements of the system are offered and maintained by different actors who follow their own very specific and dedicated agendas. Because each element of the overall system needs substantial investment and the acceptance of customers remains uncertain, companies feel tempted to wait until the whole system is in place but avoiding to initiate the first step which is commonly referred to as the chicken-egg problem (Schwoon 2008). Taking the first step implies taking on an additional risk due to the uncertainty of whether and when other elements of the system will follow.

Currently fuel cells remain expensive substitutes of internal combustion engines although it is assumed that with an increase of the scale of production, the cost will decrease resulting in economies of scale which are driven by learning effects and the sheer number of units manufactured using the initial equipment in the first place. Also the necessary manufacturing equipment will improve with the increasing number of units produced, as will the logistical efforts related to the supply chain (Schwoon 2008). Considering the supply chain, it is shown that there are a rather high number of component suppliers with comparable small numbers of units produced, which is still quite atypical for the industry value chain. However the assumption is that also in the supply chain competition will force suppliers who are currently active in the field to merge in order to achieve economies of scale and eventually lower the unit cost. Learning effects and economies of scale can be expected to contribute strongly to decreasing the cost especially in the automotive industry because of the naturally developed industry structure, which is characterized by strong OEMs and a tier supply system. System suppliers provide their solutions to more than one OEM, which enhances the learning effects of these companies. Furthermore tier 1 suppliers purchase components from a variety of tier 2 suppliers and further down the value chain which results in positive network effects (Schwoon 2008).

Particularily in the automotive supplier industries, margins are rather low which turns out to be a factor favouring marginal innovations and incremental improvements over radical disruptive innovations, which require substantial investment and carry significant risks and uncertainty of amortization (van den Hoed 2007). The latter is even more prominent and important for component suppliers who often face difficulties financing innovation with a long pay off time period. In this regard it can be assumed that the automotive industry and the supplier industry have little interest in promoting radical technological changes. It follows that for such changes to happen, the appearance of new entries, the emergence of external shocks, unexpected performance improvements of technology, sudden changes in the markets and market environments or a shift in the industry competition are required (van den Hoed 2007; van Bree et al. 2010).

Customers often consider the automotive industry as a highly innovation-intensive industry. This perception is only partially true. Due to the complexity, margins and risks, the industry tends more towards marginal innovations instead of radical breakthroughs (van den Hoed 2007). This refers not only to the initial technology, e.g., the fuel cell in this case, and the related final product specific interfaces of the technology but also to assembly lines and the broader logistical solutions in this context. In this respect another reasonable barrier for disruptive innovation in the field of car engines is the high capital intensive nature of the industry (Zapata and Nieuwenhui 2010). This is for production only, however over the full product life cycle it becomes ever more complex and capital intensive. Mainly due to the high capital intensity of the car industry, the amortization of investments is also a crucial dimension to consider in decision making, e.g., the assessment of an alternative technology is not limited to the new technology's features but also involves a financial assessment of the investments that have been made for the existing technologies, including capital cost for the equipment needed to embed technology in applications, hence products. Capital costs include expenditures for equipment and opportunity cost, as well as the investment in skills and in the labor force, which are essential to operate an equipment industry (Zapata and Nieuwenhui 2010). Although it is predicted that the cost advantage of internal combustion engines will diminish with larger volumes of fuel cells manufactured due to learning effects and economies of scale (Schwoon 2008), the broad diffusion of fuel cells also depends on transitions in the overall energy system, e.g., the availability of respective fuels and a fuel transport system among others (Turto 2006).

The environmental and the societal dimension of energy supply are closely interconnected (Dincer 2008). The environmental impact of energy generation is largely determined by the choice and use of the energy source, by the transmission of energy and by the effects which occur with the use of the energy that has been generated, in other words, the flow of generation, transmission and application. Society is typically unaware of the flow of energy as described; in general, the population's understanding of energy is that it is available as needed.

The challenges from energy supply involve a broad range of environmental challenges, namely air pollution, water pollution, solid waste, pollutants and eco-system degradation, and these problems extend over ever-wider areas (Dincer 2008). Because these types of pollution occur rather slowly, the population has difficulties in recognizing them and assessing their real impact. It seems a common social problem that the consequences of respective actions are not present in citizens' minds. Rather this is true in the case of unusual events like accident disruptions in the energy flow when the population shows increasing awareness of the environmental aspect of the latter.

3.3   Stakeholders' Hidden Agendas Analysis

Detecting and understanding stakeholders' interests and strategies requires the systematic analysis of the surrounding factors, e.g. social, technological, economic, environmental and political factors (STEEP). The STEEP analysis has become a widespread analytical concept that is used for analyzing the determinants of current and expected potential technology and innovation diffusion. It is recommended that the general characteristics of each of the five major dimensions be elaborated upon first and that the stakeholders' attitudes towards these characteristics be derived in a second step. The initial characteristics of each dimension are enriched by more general assumptions for the potential drivers which determine the characteristics currently and the resulting possible impact. This is then the basis for the specific stakeholder characteristics.

3.3.1 Social Perspective

From the social perspective, it shows that the public attitude towards hydrogen is considered to be one of the key factors towards the transition to fuel cell-powered cars. In the public perception, the availability, accessibility as well as aesthetics and convenience are the predominant features. The main issues are still the actual availability of fuel cell equipped vehicles together with the local availability and national coverage of fuel stations for long distance travel. Besides the fuel network, the maintenance infrastructure and the associated repair frequencies and costs are important in the public's perception, which drives the attitude toward the vehicles. Moreover the initial purchase cost is one of the determinants of the selection of a vehicle but it is still expected that fuel cell powered electrical vehicles (FCEVs) will be significantly more expensive than conventional vehicles and re-fuelling will be limited to a low number of locations. Aesthetics and convenience of vehicles are major determinants for public acceptance of the technology used for mobility of society, although the public attitude towards hydrogen is considered one of the key factors towards the transition to fuel cell powered cars. Thus FCEVs can be presented with a new image by combining sleek design and technology, which may be used to create a new fashion.

Given the recent development in which the automotive industry is converging more strongly with communications, electronics, and photography in the sense of integration and inter-operability, products with information, communication, multimedia systems and social networking technologies with large touch screens will also attract a number of users. Also the mass media are an important channel to promote FCEVs because it has a strong impact on public opinions. It has been frequently observed that media tends to report more on accidents and failures of technologies instead of success stories. Here the impact of different media channels on consumer attitudes and behaviour could be given special attention. Furthermore society should be convinced about the safety, security, and reliability of FCEVs, without any negative impacts on public and individual health.

3.3.2 Technological Perspective

These factors focus on rates of technological progress, pace of diffusion of innovations, problems and risks associated with technology such as security and health problems. Among the most frequently cited challenges is the reliability of fuel, e.g. centralized or decentralized fuel production and the respective infrastructure for fuel distribution and the appropriateness of the existing infrastructure for upgrading to the respective fuel distribution. Also the overall energy balance sheet of fuel production for FCEVs causes concerns among stakeholders.

Availability of equipment is an issue that comes up and requires technological progress. This relates to fuel stations and the appropriate network development, as well as to matters regarding the storage of spare parts related to fuel cell powered cars, namely the transport and actual storage of single parts, components or systems and the maintenance and repair infrastructure, e.g., physical investments in repair equipment and the training of operating staff.

Recent technological progress has the potential to impact the diffusion of technology. Mostly these advances were made in infrastructure development. It is expected that new ways of extracting hydrogen and mobile hydrogen refilling stations can be used to provide further access to hydrogen in remote or congested areas, or when a likely power cut starts effecting supply. The use of lightweight carbon fibre composite tanks for the high pressure bulk transportation of hydrogen, for mobile hydrogen fuelling station applications, and for portable self-contained hydrogen fuelling units is becoming widespread and transportable compressed hydrogen units can be custom-designed to meet customer needs, including transport trailers, mobile fuellers, and portable filling stations.

3.3.3 Economic Perspective

Levels and distribution of economic growth, industrial structures, competition and competitiveness, markets and financial issues are significant drivers of technology diffusion. From the demand perspective, the price of the end vehicle together with performance characteristics, such as the range, hydrogen consumption and overall life cycle costs are especially important. Life cycle costs also involve insurance premiums that vehicle owners have to pay and which are currently uncertain. Also it should be noted that the consumer attitudes in terms of different user segments and the size of each group vary. Initial investments in the infrastructure and remaining technological challenges are not likely to pay off earlier than the late 2020s, still business cases should include the first mover commercial advantage. It seems plausible to assume that the first and immediate clients for FCEV will be commercial fleets which also support the leverage of learning effects and cost reductions.

The nearest-term application for fuel cells seems to be lift trucks (forklifts). Several industrial truck companies have announced commercial fuel cell products that can replace battery-powered forklifts. These have been extensively tested and are available for commercial purchase today to be used in production plants, logistics and airports. A possible "electron economy" may replace the "hydrogen economy". In the "electron economy", most energy would be distributed with the highest efficiency by electricity and the shortest route in existing infrastructure could be taken. The efficiency of the "electron economy" is not affected by any wasteful conversions from physical to chemical and from chemical to physical energy. With the "electron economy", attention could be quickly turned to energy storage technologies and an upgrade to smart grids.

3.3.4 Environmental Perspective

The obvious environmental impact is one of the key arguments for supporting or contesting the transition process to hydrogen fuel cell cars. The positive environmental impacts mainly result from zero emissions from the technology and the use of widely available natural gas and existing distribution pipelines to create hydrogen for on-site fuelling. Eventually FCEVs will contribute significantly to improved air quality and a reduction in noise pollution from traffic compared with conventional vehicles powered by conventional engines. Breakthroughs in electric power storage occurred within a decade involving storage, fuel cells and new chemicals and materials including nanotechnology applications have the potential to even increase the environmental impact of FCEVs. Also highly volatile corn prices driven by a bad harvest could hurt corn ethanol producers, which are suffering from a saturated market for ethanol. This may allow hydrogen to take off faster than expected as an alternative energy source.

3.3.5   Political Perspective

Political factors involve dominant political viewpoints or parties, political (in) stability, regulatory roles and actions of governments, political action and lobbying by non-state actors. The diversification of the energy supply through hydrogen helps to reduce reliance on fossil fuels for transport which are imported products in many countries and increase energy security in energy importing countries. Furthermore the local production of hydrogen can also provide more of the process inputs to be produced locally by reducing the dependency on external energy markets. This leads to the creation of political incentives in order to promote the diffusion and acceptance of FCEVs, for example national pricing systems for hydrogen and tax exemptions planned for hydrogen vehicles. Supporting the diffusion of FCEVs governments are likely to introduce large scale public procurement programmes, government-backed zero-interest mortgage plans for hydrogen cars and massive transition of public transport vehicles and large fleets to FCEVs.

3.4   Conclusion

All relevant dimensions taken together show a clear picture of the existing and potential for, as well as the articulated and hidden arguments against a technology, e.g., the FCEV. Some arguments appear obvious but there remain major obstacles which are not considered in the public debate and presumably are not included in strategic planning and thinking of the actors. It also appears important to note that stakeholders have different influences on the respective decision making and the power to drive the mindsets of decision makers. Accordingly the arguments raised by stakeholder groups potentially prove influential to decision makers when it comes to introduction of technologies and measures supporting technology diffusion (Table 3.1). It has been observed that so far in case of fuel cell technologies

Table 3.1 Stakeholder arguments—summary

Stakeholder	Influence	Power	Argumentation strategies
Social			
• Traditional car owner	↗	↗	• Misses typical car features • Reluctant towards noiseless drive
• Young generation	↑	→	• Wish to differentiate from traditional drivers • Limited experience with infrastructure
• Car owner association	↑	↑	• Adverse attitudes, mainly dominated by traditional driver • Point on noise, danger of fuel supply, need to train traditional driver to adjust
Technological			
• Producers of solid oxide fuel cell (SOFC), phosphoric acid fuel cell (PAFC), molten carbonate fuel cell • (MCFC)	→	↗	• Similar application fields for fuel cells or at least potentially similar fields • Might point to dangers and environmental issues of membranes used in proton exchange membrane (PEM) fuel cells
• Infrastructure supplier	↑	↑	• Decentralized infrastructure needs to be built – investment cost • Existing infrastructure reshaped for fuel transport – opportunity cost
• Fuel producer (gasoline)	↗	↑	• Consequences of fall in demand for gasoline – refinery closures, job losses, impact on petrochemical industry
Economic			
• Technology follower	↗	↗	• Technological leadership concentrated in Asia (Japan, South Korea), Europeans lagging, oppose with lobby work
• Petrochemical industry	↑	↑	• Job losses due to either refinery closure or high investment in new equipment
• Repair and maintenance industry	↗	↗	• Significant investment in equipment • No competences in new technologies, reluctant to accept dual system
Environmental			
• Laws, legal regulations	↑	↑	• Specially important for environment, health and safety issues

(continued)

Table 3.1 (continued)

Stakeholder	Influence	Power	Argumentation strategies
• Environmental groups	↗	↗	• Long-term hydrogen impact not known
• Health, safety groups	↑	↑	• Unknown reliability of new standards and technologies, potential negative impact on safety and health of workers in all domains
Political			
• Municipalities	↑	↑	• Responsible for infrastructural decisions
• Regional	↑	↑	• Financial incentives for municipalities, regional standards, complementarities of standards between regions
• Federal	↗	↗	• Initiator and promoter role but less implementation power

Legend: influence, power: →—low; ↑—strong; ↗—medium
Source: National Research University Higher School of Economics

many of these arguments were not or only partially reflected in the technology and innovation strategy development although they appear almost equally important for a targeted and effective communication of the technologies to customers.

In summary it can be concluded that the FCEV is an option for transportation, but it presumably has a short life cycle if stakeholders beyond transport are not considered. Also it seems that FCEV is fulfilling a bridging role from the hydrogen century ahead towards the electron century, which is expected to come in the future. Given the accelerating speed of science and technology development, some are already raising concerns about investments in the hydrogen century as it is believed that the transportation of the future will be decided by the next standard, e.g., hydrogen vs. pure electron. Also there is an indication that stakeholders, namely users' attitudes, might potentially change if investment in standards and users education are going together. Standard setting at the current stage may appear to be a safeguard for decades of technology survival and is a precondition for justifying the respective substantial investments.

Currently transport-related roadmaps show weaknesses in considering the agendas of all actual and potential stakeholders. Market roadmaps commonly assume only a modest change in customers' behaviors and focus strongly on competing products and technologies but less on the actual attitudes of users and social agendas and values, etc., which means that mostly the 'hidden' stakeholders, who become obvious if one analyses the systemic impact of FCEVs, are neglected. Still as experiences show, the pace of technology and innovation diffusion, are at least partially determined by the early involvement of all stakeholders. Involvement does not necessarily mean the active inclusion of the stakeholders for adoption of technologies. Rather it is more important to learn the actual agendas and to elaborate on the hidden agendas of stakeholders, which allows for the identification of existing and potential obstacles for the broad application and diffusion of a technology. From a purely technical point of view, the main challenges of fuel cells related to the engine, recovery, storage and transfer of hydrogen can be considered solved in principle. Obviously the main hurdle that remains is the integration into the overall system, e.g., the FCEV and acceptance by the market. This includes the technical and also the manufacturing integration of the fuel cell itself but as well as the FCEV, which requires a new logistics system, hence a renewal of the supply chain in the first instance. This also goes along with renewal of the maintenance and recycling infrastructure which are separate dedicated ecosystems. Also market acceptance from customer perspective needs certification procedures imposed by governmental bodies in the first instance and the trust of customers in the second instance.

Regulations and certifications are especially important in the automobile industry. Moreover, even if OEMs have obtained all certificates, what the media is reporting is very important. The experience of Asian OEMs aiming to enter the European car markets gives a substantial indication of the power of media in the diffusion of new products even though by that time the products were reasonably similar. However, in this case, Asian OEMs had to struggle with market entry and penetration for a while. In this respect, it can be argued that media strongly influences public attitudes towards technologies and therefore towards products. Therefore, the media's role as a stakeholder in the diffusion of FCEVs is more important than was previously assumed. Also in this respect, social media, namely internet-based, has its own influence. One might assume that internet media mainly is used for information collection and experience exchanges. The latter naturally becomes crucial for the formation of opinions by customers especially with respect to vehicles.

In the short term stakeholders may raise skepticism, in which the municipalities already claim and will probably continue to claim that they cannot guarantee the safe operation of hydrogen supply system, which is largely due to a lack in the certification and qualification infrastructure. The petrochemical industry expertly argues that fuel cells are reasonable to use for vehicles, but also raises concerns about the downstream industry which might suffer as a result of decreasing demand for gasoline, which is a common refining product and this might place pressure on the industry due to the emerging need to upgrade existing facilities and equipment. This need for renewal is based on the nature of commonly applied processes as a result of which fuel is produced as one product among others. However, changing the refining processes in favor of less or no fuel (gasoline) production needs different equipment.

Until recently a maintenance infrastructure for vehicles (garages, workshops) is in place, which was dedicated to combustion engines with all the necessary equipment available. Changing to FCEVs, however, requires an almost full scale conversion of the existing maintenance infrastructure towards a combined combustion engine and fuel cell maintenance equipment. This can only be done with considerable financial investment in technical equipment and also in the training of maintenance professionals in the short term. Presumably the owners of the existing infrastructure would be reluctant and skeptical about such significant investments as long as substantial demand remained unpredictable.

There are also mid-term threats for the fuel cell, namely the inherent danger that society maybe sceptical about the safety, security, and reliability of such a highly flammable and explosive fuel as hydrogen in cars. The rapid diffusion of hybrids and the progress in battery technologies may delay the commercial implementation of FCEVs, e.g., a possible "electron economy" may replace the "hydrogen economy". With the electron economy, attention could be quickly turned to the energy storage technologies and an upgrade to smart grids. The current enthusiasm for electrical vehicles can also be traced back to governments' announcements but one might speculate that after a couple of serious accidents, the governments may withdraw their support for these (likewise the case of Fukushima discouraged some governments from the use of nuclear power plants).

In conclusion, the stakeholder analysis is a widespread approach used for technology and innovation diffusion strategy building. A broad range of social, technological, economic, environmental and political issues show a considerable impact on the diffusion of technologies as shown in the sample case of fuel cells.

Acknowledgement The chapter was prepared within the framework of the Basic Research Programme at the National Research University Higher School of Economics (HSE) and supported within the framework of the subsidy granted to the HSE by the Government of the Russian Federation for the implementation of the Global Competitiveness Programme.

References

Afgan NH, Carvalho MG (2004) Sustainability assessment of hydrogen energy systems. Int J Hydrogen Energy 2004:1327–1342
Afgan NH, Al-Gobaisi D, Carvalho MG, Cumoc M (1998) Sustainable energy development. Renew Sustain Energy Rev 2:235–286
Baretto L, Makihira A, Riahi K (2003) The hydrogen economy in the 21st century: a sustainable development scenario. Int J Hydrogen Energy 28:267–284
Brown MM (2003) Technology diffusion and the "knowledge barrier": the dilemma of stakeholder participation. Publ Perform Manag Rev 26(4):345–359
Collantes GO (2007) Incorporating stakeholders' perspectives into models of new technology diffusion: the case of fuel-cell vehicles. Technol Forecast Social Change 74:267–280
Dincer I (2008) Hydrogen and fuel cell technologies for sustainable future. Jordan J Mech Ind Eng 2(1):1–14
Dincer I, Rosen MA (1998) A worldwide perspective on energy, environment and sustainable development. Int J Energy Res 22(15):1305–1321
Dincer I, Rosen MA (2005) Thermodynamic aspects of renewable and sustainable development. Renew Sustain Energy Rev 9(2):169–189
Friedman A, Miles S (2006) Stakeholders: theory and practice. Oxford University Press, Oxford
Geroski PA (2000) Models of technology diffusion. Res Pol 29:603–625
Hammond GP (2004) Towards sustainability: energy efficiency, thermodynamic analysis, and the 'two cultures'. Energy Policy 32(16):1789–1798

Hart D (2000) Sustainable energy conversion: fuel cells-the competitive options. J Power Sources 86:23–27

Hopwood B, Mellor M, O'Brien G (2005) Sustainable development: mapping different approaches. Sustain Dev 13:38–52

Hui SCM (1997) From renewable energy to sustainability: the challenge for Hong Kong. POLMET '97 Conference, Hong Kong

Jepsen AL, Eskerod P (2009) Stakeholder analysis in projects: challenges in using current guidelines in the real world. Int J Project Manag 27:335–343

Keles D, Wietschel M, Moest D, Rentz O (2008) Market penetration of fuel cell vehicles—analysis based on agent behaviour. Int J Hydrogen Energy 33:4444–4455

Kutsenko E, Meissner D (2013) Key features of the first phase of the national cluster programme in Russia. In: Working papers by NRU Higher School of Economics. Series WP BRP "Science, Technology and Innovation". No. 11/STI/2013

McGowan JG (1990) Large-scale solar/wind electrical production systems-predictions for the 21st century. In: Tester JW, Wood DO, Ferran NA (eds) Energy and the environment in the 21st century. MIT, Massachusetts

Meissner D (2014) Approaches for developing national STI strategies. STI Pol Rev 5(1):34–56

Meissner D, Roud V, Cervantes M (2013) Innovation policy or policy for innovation? In search of the optimal solution for policy approach and organisation. In: Meissner D, Gokhberg L, Sokolov A (eds) Science, technology and innovation policy for the future—potentials and limits of foresight studies. Springer, Heidelberg, New York, Dordrecht, London, pp 247–255

Midilli A, Ay M, Dincer I, Rosen MA (2005a) On hydrogen and hydrogen energy strategies-I: current status and needs. Renew Sustain Energy Rev 9(3):291–307

Midilli A, Ay M, Dincer I, Rosen MA (2005b) On hydrogen and hydrogen energy strategies-II: future projections affecting global stability and unrest. Renew Sustain Energy Rev 9 (3):309–323

Pehnt M (2003) Assessing future energy and transport systems: the case of fuel cells. Part I: methodological aspects. Int J Life Cycle Assessment 8(5):283–289

Proskuryakova L, Meissner D, Rudnik P (2014) A policy perspective on the Russian technology platforms. In: Working papers by NRU Higher School of Economics. Series WP BRP "Science, Technology and Innovation". No. 26/STI/2014

Proskuryakova L, Meissner D, Rudnik P (2015a) The use of technology platforms as a policy tool to address research challenges and technology transfer. J Technol Transfer. Available online 10.1007/s10961-014-9373-8

Proskuryakova L, Meissner D, Rudnik P (2015b) Technology platforms as science, technology and innovation policy instruments: learnings from industrial technology platforms. STI Pol Rev 6 (1):70–84

Reed MS, Graves A, Dandy N, Posthumus H, Hubacek K, Morris J, Prell C, Quinn CJ, Stringe LC (2009) Who's in and why? A typology of stakeholder analysis methods for natural resource management. J Environ Manage 90:1933–1949

Rodriguez M, Paredes F (2015) Technological landscape and collaborations in hybrid vehicles industry. Foresight-Russia 9(2):6–21

Schibany A, Reiner C (2014) Can basic research provide a way out of economic stagnation? Foresight-Russia 8(4):54–63

Schwoon M (2008) Learning by doing, learning spillovers and the diffusion of fuel cell vehicles. Simulation Modell Pract Theory 16:1463–1476

Simachev Y, Kuzyk M, Kuznetsov B, Pogrebnyak E (2014) Russia heading towards a new technology-industrial policy: exciting prospects and fatal traps. Foresight-Russia 8(4):6–23 (in Russian)

Turto H (2006) Sustainable global automobile transport in the 21st century: an integrated scenario analysis. Technol Forecast Social Change 73:607–629

Usha Rao K, Kishore VVN (2010) A review of technology diffusion models with special reference to renewable energy technologies. Renew Sustain Energy Rev 14:1070–1078 van Bree B, Verbong GPJ, Kramer GJ (2010) A multi-level perspective on the introduction of hydrogen and battery-electric vehicles. Technol Forecast Social Change 77:529–540

van den Hoed R (2007) Sources of radical technological innovation: the emergence of fuel cell technology in the automotive industry. J Cleaner Prod 15:1014–1021

Vishnevskiy K, Karasev O, Meissner D (2015) Integrated roadmaps and corporate foresight as tools of innovation management: the case of Russian companies. Technol Forecast Social Change 90(B):433–443

Zapata C, Nieuwenhui P (2010) Exploring innovation in the automotive industry: new technologies for cleaner cars. J Cleaner Prod 18:14–20

Integrating Foresight with Corporate Planning

4

Jonathan D. Linton and Steven T. Walsh

4.1 Introduction

This chapter proposes and illustrates an approach for organizations to respond to foresight and roadmapping exercises. While foresight and roadmapping activities are critical for the management and stakeholders of organizations to understand what might be required in the future, in itself the knowledge does not bring the organization any closer to the future anticipated state(s). Consequently, this article provides a framework and approach to determining an organization's current state. Migration from current to possible future state(s) is achieved by recognizing the existing gaps and by closing these gaps through some combination of: development of internal capabilities, acquisition/merger with other organizations, outsourcing, and/or partnering and leveraging other elements of the supply chain. We focus on the determination of current state and then offer an illustration of comparison to future state and identifying the gaps that may need to be overcome.

A number of decomposition models have been put forth to obtain an understanding of what is required for success in a desired product market (Adler 1989; Fusfeld 1970, 1978; Walsh and Linton 2001, 2011). These models have roots in several different streams of literature: the importance of technology and innovation (Solow 1957; Schumpeter 1934), Resources Based Theory (Barney 1991), and Competence-Based Theory (Bitindo and Frohman 1981; Prahalad and Hamel

J.D. Linton (✉)
Telfer School of Management, University of Ottawa, DMS 6108, 55 Laurier Street, Ottawa, ON K1N 6N1, Canada

Institute for Statistical Studies and Economics of Knowledge, National Research University Higher School of Economics, 20 Myasnitskaya Street, 101000 Moscow, Russia
e-mail: Linton@UOttawa.ca

S.T. Walsh
Anderson School, University of New Mexico, 1924 Las Lomas NE, Albuquerque, NM 87131, USA

© Springer International Publishing Switzerland 2016
L. Gokhberg et al. (eds.), *Deploying Foresight for Policy and Strategy Makers*, Science, Technology and Innovation Studies, DOI 10.1007/978-3-319-25628-3_4

1990, 1991). While much literature focuses on the opportunity (Bhave 1994; Alvarez and Busenitz 2001; Park 2005), we focus on the nature of an existing firm. The firm's current abilities are critical for its future directions as they govern the *tracks in the sand* created by a firm (Mintzberg 2007). More specifically, it is necessary to consider both *managerial capabilities* and *technological competencies* (Marino 1996). While arguments are offered that the abilities are frozen in place and to some extent cannot be changed due to the existence of core rigidities (Leonard-Barton 1992), we consider what exists and what is needed and leave questions of organizational flexibility and change management to others. Having offered an overview of the Strategy Technology Firm-Fit Audit (Walsh and Linton 2011), the different elements are briefly introduced and, the process for utilizing the audit in conjunction with roadmaps and/or other foresight exercises is offered, this is followed by the illustration of clean room development for the electronics and semiconductor industry which is pursued by concluding notes.

4.2 A Framework for Understanding Available and Needed Abilities

The elements—*managerial capabilities* and *technical competencies*—are discussed briefly as a detailed development and explanation can be found in Walsh and Linton (2011).

4.2.1 Managerial Capabilities

Management skills and routines differ depending on context. Utilizing appropriate skills in the incorrect context can be problematic, so it is important to clearly understand the context in which one is operating so the appropriate skills can be utilized. These appropriate managerial skills can be broken down into the following categories: offering type, type of physical product, type of service product, value emphasis, complexity, and nature of technological change and innovation.

The offering type indicates whether one is selling a physical product, a service or a combination of the two or after-sales service. Traditionally, the sale of ownership of physical products has been the focus of companies. In some cases physical products are not actually sold, but the use of their associated benefits is provided instead (Michaelis and Coates 1994). Alternatively, a product does not involve the transfer or direct use of a physical good, but is focused on the provision of some sort of service. The management of product and services differ, so it is important to recognize this as a sudden change from one category to another will present new and unfamiliar challenges and concerns. Many companies that have traditionally focused on the sale of physical products, have recognized that the provision of associated services is very profitable (Baumgartner and Wise 1999). These after-sales services present different management challenges than either a pure physical product or a pure service as the services are clearly associated to the physical product and benefit from knowledge specific to the product. Hence after-sales service is a separate category and offers insights into the sale of product-benefit (a service) or the provision of services—training, maintenance, product optimization—after the ownership of a product has transferred.

Physical products can be divided into two categories: *materials* and *fabrication and assembly*. Products based on *Fabrication and Assembly* of components include: electronics, automotive, and aerospace. These products are managed differently than *Services* or *Materials*-based products. For example, these sorts of products typically involve intense product innovation being followed by process innovation (Utterback 1994). While *Materials*-based products differ as they are based on flows and processes. Consequently, they lack a unit form. This category includes: engineered materials, chemicals, and food. The process-based nature of the products makes management of the technology unique. For *Materials*: product and process innovation occur simultaneously. When dealing with a physical product, it is important to denote whether one is dealing with *fabrication and assembly* or *materials*. This is a critical consideration as changes in future manufacturing orientation are likely to move many products from one category to the other. For example, traditional electro-mechanical sensors relied on *assembly* of components. While, the change to MEMS-based production technologies has resulted in a *materials*-based semiconductor like process. Self-assembling systems (nanotechnology) and 3D Printing are two futuristic process technologies that are likely to move the skill sets required for many products that currently rely on *fabrication and assembly* to *materials*-based management practices.

Service products should be managed differently depending on whether they are *Knowledge Embedded, Based*, or *Extracted*. In *Knowledge Embedded* services the service has been constructed utilizing a great deal of knowledge so a sophisticated product can be delivered with little effort. A car wash or fast food restaurant is an example of this sort of service as tremendous effort is put into the design of the service so that it can offer a consistent product with very little effort. Consequently, the service provider needs very little skill to offer a product of high quality and consistency. *Knowledge Based* differs greatly as the important abilities, understanding and knowledge resides in the service provider. The value of the service is a function of the abilities and experience of the service provider. Professional services (also referred to as KIBS—Knowledge Intensive Business Services) are the most obvious examples. Physicians, attorneys, accountants are examples of service providers that offer value based on their personal abilities and understanding of the needs of the client. Supporting systems and equipment, while useful, are clearly subordinate relative to the personal skill(s) of the service provider. Finally, *Knowledge Extracted* involves the user or customer interacting or co-creating the required service with a process that has been carefully designed, so that most users are comfortable and capable of operating the service without aid. An example of this is the Automated Teller Machine (ATM)—a device that has been carefully designed to provide a wide range of valuable services in a manner that most user/consumers can access unaided. Having considered physical products and services, it is worth making note of after sales service.

Many firms clearly are producers of physical products. Their managerial skills and systems are set up for this purpose. These firms clearly differ from services firms in terms of the presence and absence of skills. *After-Sales Service* is especially worth mentioning for physical product firms, but is also relevant for service firms. Some firms sell a physical product and then expect to never have any interaction—other than future product sales—with the customer. While other firms offer a variety of after-sales services. The presence or absence of after-sales service offerings is worth noting as it identifies the presence/absence of important types of knowledge, skills and systems. *After-sales services* are typically high in margin, have a multiplier effect on the revenue associated with a single product sale and serve to create switching costs by creating further bonds between firm and customer (Baumgartner and Wise 1999). After-sales services include activities such as: financing, parts and maintenance services, training, and equipment performance agreements. If after-sales services are provided it is worth enumerating the types of services provided to better understand what sort of capabilities are in house and which are lacking. The presence of these services is expected to increase in importance as customers and firms, increasingly, see suppliers as strategic partners in a supply chain as opposed to single transaction price-based commodity providers. The increasing consideration of sustainability also is a driver for the increasing importance of after-sales services to extract the maximum value and benefit from the purchase of physical product. Having considered the physical, service, and after sales service form of products; we now consider how products offer value.

Value emphasis—can be based either on *operations* or *technology development*. If the value emphasis is based on *operations* the focus is on process and production. In which case research activities and intellectual property is geared towards how the product is made—not the actual product. While product design, features, and positioning are still critical; the value is a based on the ability to produce the product at a cost much lower than the market prices and associated perceptions of customer value. Manufacturers of bulk chemicals and materials are typical examples of this. *Technology development* firms focus on the development of products. In some cases these firms outsource all of manufacturing and production as they do not create value through the production of a product, but by the product in itself. While Apple Inc. is a classical example as they focus on development of technology and design of product, while manufacturing is relatively traditional and simple and can be outsourced to a third-party. Integrators such as Dell and Cisco focus on product design. There is very little physical transformation of product conducted by either company. Both companies rely on their supply chain to physically transform raw materials into components, modules and systems that can be used as part of their product. Having considered how value is added, the importance of the level of product/process complexity is considered.

Complexity—differs for products as a function of the number of components or processes involved. At its most simple, a product can be made with very few processes or components. As the number of processes or the number of components increases the management complexity increases. In the case of increased components, management skills relating to such things as: vendor selection and management, inventory control, logistics, and vendor quality control become increasingly important. Also important, but for different reasons, is increasing complexity due to a larger number of manufacturing processes. The management of a large number of manufacturing processes requires managerial skills in areas such as process quality control, process set-up and change over, continuous improvement, maintenance, and capacity planning. While none of these skills are particularly difficult to obtain and maintain, there is a clear difference in skill sets and a lack of recognition of the need to possess and/or develop these skills is problematic. As the number of components or processes increase, the importance of specific managerial skills increasingly are required for not only sustainable competitive advantage, but for any level of competence and competitiveness. Hence, it is important to distinguish a firm or product as having, either: (1) few components and/or processes, (2) moderate number of components and/or processes, or (3) large number of components or processes. If future requirements are for a smaller number of components or processes, this is an easy adjustment as the existing managerial skills will outstrip future requirements. However if future requirements involve a clear increase in the number of components or processes, it is quite possible that current managerial skills and systems are insufficient. Therefore when the need to evolve to deal with larger numbers of components and or processes is identified as a possibility, an assessment of skills gaps and a plan for closing these gaps is required.

Pace of technological change consists of two separate parts: *technology maturity* and the *nature of innovation*. The *technology maturity* is a measure of how well understood and developed a technology is. This category can pertain to either a process or product technology. In many cases it relates to both simultaneously. If the maturity level of a technology is high, it is not of particular interest, concern and potential advantage as the management, control, and use of the technology are well understood. While having a strong understanding of a technology is comforting and risk reducing, the technology ceases to be an important contributor to future value. Additional value through new insights and opportunity for innovation is usually very low. For high *technology maturity* products the absence of potential upside value through further understanding and innovation, limits the need for research, development and flexible systems. A very stable mechanistic environment is suitable for working with such technologies (Burns and Stalker 1961; Morgan 2006). As *technology maturity* declines—it becomes moderate or even low—the opportunity and *need* for increasing level of understanding to obtain consistent and better results increases. Consequently, the presence of technical skills, experimentation and the ability to evolve product and process to take advantage of now knowledge is increasingly important. For more detail on the sort of learning and associated skills required as one moves from a completely immature technology to a very mature technology see Gomory (1989) and Bohn (1993). In addition to the maturity of the technology, the type of innovation is critical.

While innovation type has some relation to the maturity of technology, this is often not the case. Consequently, it is worth considering them together, but separately. The seminal work considering innovation type is that of Abernathy and Clark (1985) who break innovation into four separate types: *regular, niche, revolutionary* and *architectural*. Most common is *regular* (also referred to as *incremental or evolutionary) innovation* which involves improvements that build on the existing technological knowledge and assets and further builds the value of a firm's technological and production skills and the existing product base. *Revolutionary (or radical) innovation* builds on a different base of knowledge and technology. Consequently, it destroys existing technical competencies and in the process captures existing markets. The replacement of analog electronics with digital electronics is one example. *Niche* innovation involves the use of the same technological base in a different way to modify and/or destroy existing market connections. The role the internet has played in eliminating (or reducing the value of) traditional distribution chains and middlemen is an example. Finally, *architectural innovation* sees a replacement/destruction of both the original technological base and the associated market connection. Both of which are replaced by something new and clearly superior in one or more directions. The replacement of banking services with the use of the cell phone as a way of storing and transferring funds between people, organizations and locations is the use of both a different technological base and set of market connections. Having considered the four major types of innovation, it is also worth considering the dichotomy of *market pull* and *technology push* (Munro and Noori 1988; Veugelers and Cassiman 1999; Walsh et al. 2002). With *market pull* the organization looks to market needs to determine what innovations to develop. This approach is consistent with the concepts of *Voice of the Customer* (Sullivan 1986; Thompson 1997; Matzler and Hinterhuber 1998) and with the lead-user research (von Hippel 1976, 1986, 1988). Technology push relies on the development of a technology and the identification of characteristics of the technology that lead to a unique value proposition to one or more markets. Firms that are based on the presence of a core competence (Prahalad and Hamel 1990) are *technology push* firms.

Having provided a summary of the different types of managerial skills that should be considered when looking at the capabilities of a firm or the needs associated with a specific opportunity, attention is now transferred to the technical aspects—that is technological competencies.

4.2.2 Technical Competencies

While all branches of science interact with each other, it is the practice of our system of education and specialization to place clear divisions between fields. It is quite possible that the increasingly apparent convergence of fields will change this in the future. However, at the present time scientific disciplines are clearly defined and separated. For more specific and specialized areas there are a myriad of sub-fields. So many sub-fields are present that one must consider a specific industry, product or firm to clearly state relevant sub-fields in the form of a list of manageable size. Consequently, for scientific disciplines a full list can be offered for any assessment. While, careful consideration must be given to determine the relevant subfields. As our focus is on applied sciences and technologies, both science and engineering skills sets are concerned. These discipline-based skills are referred to as *generic technological skills*. To date we have found that *generic technological skills* can be covered in most cases by the following list: *biology, chemistry, civil engineering, computer science, electrical engineering, materials science/engineering, mechanical engineering* and *nanotechnology*. Of course additional *generic technological skills* can be added on if the analyst feels it helps to better assess an area of interest.

Having identified which of the *generic technological skills* are relevant it is important to determine what *specific engineering skills* are needed within each of the *generic technological skill* areas. From the perspective or general management needs and entry-level recruitment, identification of the appropriate *generic technological skills* is sufficient. However, from the perspective of acquisition and maintenance of specialized skills for sales engineers, R&D personnel, process engineers, designers, procurement personnel identification and consideration of *specific engineering skills* is needed. Having identified all the needed elements to understand the needs for a firm in terms of managerial capabilities and technological competencies for competing and developing/maintaining a competitive advantage in a technologically intense area, the conglomeration of these concepts into a framework is now considered.

4.2.3 Overall Model and Its Application

The different elements discussed above are brought together as a framework that first considers the existing (and needed) management capabilities for a firm (Table 4.1) and then the existing (and needed) technological competencies (Table 4.2). Management capabilities are divided into two tiers. The first tier considers the nature of the product. That is a *physical* versus a *service* product and to further breakdown the nature of the product by considering subcategories: *fabrication and assembly, materials, knowledge embedded, knowledge based* and *knowledge extracted*. In the second tier questions relating to the nature of value generation, complexity and innovation are considered. More specifically, one considers the role of *operations* versus *technology development*, the number of component and/or processes, and various facets of innovation. The multifaceted nature of innovation results in three different categorizations being utilized to consider it: *technology maturity*, type of innovation (*regular, revolutionary, niche* or *architectural*) and whether it is driven by *market pull* or *technology push*. Having considered the managerial capabilities technological competencies are considered in the second two parts of the framework.

Technological competencies (Table 4.2) consist of two separate areas the general *generic engineering skills* that are expected to remain the same from firm-to-firm and the more specialized *specific engineering skills*. The specific skills will vary not only from firm-to-firm, product-to-product, and industry-to-industry but also over time. In the case of integrating foresight with corporate planning the recognition of *specific engineering skills* requirements evolving over time is

Table 4.1 Summary of managerial capabilities associated with strategy fit audit

Managerial capabilities	Product 1	Product 2...	Product n
Offering type			
Physical product			
Service product			
After sales service			
Differences in physical products			
Materials			
Fabrication and assembly			
Differences in service products/After sales service			
Knowledge embedded			
Knowledge based			
Knowledge extracted			
Managerial emphasis			
Operations			
Technology development			
Complexity			
Few components or process steps			
Moderate number of components or process steps			
Many components or process steps			
Technology maturity			
Low, medium or high			
Type of innovation			
Market driven (regular or niche)			
Technology push (revolutionary or architectural)			

critical. In fact, it is quite possible that *generic engineering skills* will also change over time. For example, computer science and electronics has become a critical part of many products that were mechanical only a few decades ago. Not only will this particular trend become more apparent, but the increasing importance of biology and nanotechnologies will become more apparent and the tendency towards convergence of fields will continue. A new awareness of the likely changes in *generic engineering skills* and *specific engineering skills* will have a substantial direct effect on corporate planning. The implications of these changes will often be significant for the changes in required managerial capabilities. However, these differences will be more subtle as they are indirect.

Having described the framework its application for the integration of foresight and corporate planning is now considered. The framework allows us to compare the state of an organization at the present time with the potential needs in the future. In doing so, one is able to determine to what extent that a firms current technological competencies and managerial capabilities prepare it for the future and what gaps appear to exist between current abilities and future needs.

In order to do this, one must:

Table 4.2 Summary of technological competencies associated with strategy fit audit

Technological competencies	Product 1	Product 2...	Product n
Tier III			
Generic engineering skills			
Biological			
Chemical			
Civil			
Computer Science			
Electrical			
Materials			
Mechanical			
Nanotechnologies			
Others			
Tier IV			
Specific technologies required by opportunity (some examples provided)			
Lighting technology			
Vacuum systems			
Sensor arrays			
Micromachining			

1. Fill out framework based on firm current state.
2. Fill out framework based on foresight exercise.
3. Identify gaps that exist—required in future state, but not current state.
4. Determine which gaps should be closed and what steps will be taken to close the gaps that are deemed important.

Having summarized the general process, each step is now considered in greater detail.

4.2.4 Firm: Current State

The current state of a firm is assessed through the consideration of each product or product line and its nature in terms of required managerial capabilities and technological competencies. Typically one finds that multiple product firms have a portfolio of products that for the most part draw on the same set of managerial capabilities and technical competencies. If a competence or capability is only present in a small part of the product line, it is possible that it is not well developed. However, if a competence or capability is well developed it is likely that it is entrenched throughout the organization in terms on knowledge, routines and culture. Several examples of assessments of firms are available in Walsh and Linton (2011), the readers are referred there for further details.

4.2.5   Future State

Developing a list of needed managerial capabilities and technological competencies will differ depending on the nature of the foresight activity being conducted. If an analysis involves a single roadmap or set of outcomes, it is possible to take the results and use them to assess the needed managerial capabilities and technological competencies for success in this future environment. However, if the foresight activities result in the appearance of possible alternative paths or if scenario analysis is utilized then putting together a future state map is more complex. When dealing with two or more alternative states that the future may follow, it is critical that the framework be completed for each of the possible future states. Once this is conducted, a summary framework is assembled. It should be clearly noted which of the different future states are associated with each managerial capability and technical competence. In doing so, it becomes apparent if all, most, few or a single possible future state is associated with a managerial capability or a technological competence. This provides assistance in determining the likelihood that these abilities will in fact be needed in the future. By footnoting which competencies and capabilities are associated with which of the future states are relevant, the likelihood of future importance of the ability is clearer based on the number of possible states associated with each particular item/ability. It is also apparent under what circumstances the competence or capability does not matter. That is, if a competence is associated with a single possible future state and it becomes apparent over time that the future state is of lower likelihood or lower importance the criticality of the competence declines. (The converse is of course also true.) Once a summary framework has been produced that merges the different possible future states together and this summary includes footnotes to show which states are associated to which competencies and capabilities, it is possible to consider the existing gaps.

4.2.6   Existing Gaps

It is typical in management to assume that the future is just an extension of the present. For consideration of managerial capabilities and technical competencies we can do the same to some extent. If a managerial capability or technical competency currently exists within a firm due to the existing portfolio of products, we can generally assume that these abilities will be available to the firm in the future. If there is a plan or expectation that certain products will be divested or discontinued it is important to make a note of this as it can result in the elimination of abilities that while not required at some point in the future are in fact needed shortly afterwards. The elimination of existing skills that are likely to be needed in the future is a substantial problem. Power generation with nuclear fission in many countries offers an exemplar. For several decades building of nuclear fission plants has been discontinued or downscaled. Consequently, the needed competencies and capabilities have been eliminated in many cases from corporations, governments, and educational institutions. While these competencies and capabilities may not be required in the recent past or the current time, they are likely to be needed in the future. Consequently, it is important to recognize the future need(s) and to maintain the required competencies and capabilities (as it is easier, less costly, and more flexible to maintain an existing ability than to abandon and redevelop at a later date).

Having identified if any competencies or capabilities are likely to be discontinued, we can now consider existing (and possible future) gaps between the current state and the predicted future state. Each existing or possible future gap is noted. Along with each of these possible gaps, a footnote should be offered to identify under which future state(s) each gap is expected to exist. With a list of anticipated future gaps in hand, it is time to consider the management of these gaps.

4.2.7 Gap Management

For each of the gaps identified a decision needs to be made whether an attempt will be made to fill the gap or if the management will adopt a wait-and-see approach. Immediate action is appropriate for gaps that have urgency due to their (1) importance, (2) common theme among future states, and (3) perceived difficulty to close the gap. A gap is *Important* if the presence of the gap is likely to make a firm be uncompetitive, unsuccessful or unsustainable in the future. For example, firms that were unable to adopt and utilize CFC (Chloro-Fluoro-Carbon) alternative technologies were unable to survive once the use of CFCs was banned. *Common Themes* are abilities that are required by many future states. Consequently, this ability is likely to be associated with competitive advantage of a firm in the future regardless of whether few or many of the possible future states identified do not actually turn out to be correct. Hence these gaps should be closed in a controlled and timely manner, as they are abilities that are undoubtedly required for success in the future. *Difficulty to Close the Gap* is the final critical factor. If a gap is easy to close, it is possible to postpone developing the relevant ability. However if a gap is difficult to close, then it is important to start the process early so that the ability is acquired by the time it is required.

If one or more of the three above-mentioned criteria are met by a gap—a plan to close the gap should be put in place immediately. Wait-and-see is only appropriate in cases where an ability is: of low importance, associated to only one or few a possible future states, or is easy to close. Gaps can be closed in multiple ways: developing in-house skills, hiring of key personnel, joint ventures or strategic alliances with other organizations, acquisition of other firms, or reliance/outsourcing to supply chain partners. The make/buy literature (Howells 1997; Leiblein et al. 2002) is best consulted to determine which one or set of these tactics is most suitable for a given organization and situation. Having considered the process in general a brief illustration is given to assist in clarifying the process and the value of its contribution to corporate planning.

4.3   An Illustration of the Framework: Contamination and Clean Room

As transistors—central to both semiconductor and electronics industries—continued to shrink in terms of size and the intricacy of integrated circuits increased, the yield started declining. This decline while partially counteracted by the increased density of transistors and the increased value of the more complex integrated circuits, would benefit greatly from an increase in yield that would allow for prices to be lowered and profits to be increased simultaneously. Consequently, a decision was made to try and understand the driver(s) of transistor failure so that greater yields could lead to reduces in costs, material inputs, and environmental impact. Simultaneously, increased yield offer the benefit of increased profit and capacity. This concerns affected both the associated upstream (semiconductor) and downstream (electronics) industries.

Foresight activities such as roadmapping identify the physical barriers and problems that need to be overcome. In the case of densification, one of the major causes of product failure was determined to be contamination. Consequently, a manner in which contaminants could be reduced was needed.

Elimination of contaminants using of laminar flow clean room technology was proposed by Sandia National Laboratories in response to a yield problem being suffered by one of its suppliers. As this approach was novel at the time a Sandia National Laboratory Scientist, presented his findings at a conference on contamination and manufacturing. RCA (radio electronics) and Motorola (semiconductors) were among the participants at the conference and immediately saw this innovation as a potential solution to the yield concerns they had relating to contamination. Once a decision was made to utilize this technology to hopefully partially or completely overcome their problems with contamination, the question became how could one become expert in this area. At that time, clean room equipment was not being made. Consequently, early adopters such as Sandia National Laboratories, RCA and Motorola had to manufacture their own equipment. This was done in part by working with suppliers of the critical component—HEPA filters—to have them develop and supply the needed components. Special air handling equipment was developed and manufactured by the early adopters for their own use. It is not until some years later that firms in electronics, semiconductor and other industries are able to purchase clean room equipment from suppliers.

To understand what the implications of clean room technology for the managerial capabilities and technological competencies required by a firm, the abilities needed for clean room operations were determined (Column 2, Table 4.3). The existing—for that time—managerial capabilities and technological competencies for semiconductor (Column 3, Table 4.3) and electronics (Column 4, Table 3) have been placed alongside the requirements for clean room operations. From this we can see the similarities and gap. For all the similarities—*physical product, fabrication and assembly, revolutionary and architectural innovation*, and *generic engineering skills* of *physics, chemistry, mechanical* and *materials engineering*—there was no concern regarding the firm divesting itself of an ability at a time when the

Table 4.3 Laminar flow clean room technology

Managerial capabilities Tier 1	Laminar	Semi	Electronics
Offering type			
Physical product	X	X	X
Service product			
After sales service			
Physical products			
Materials		X	
Fabrication and assembly	X		X
Service products/After sales service			
Knowledge embedded	X		
Knowledge based			
Knowledge extracted			
Managerial emphasis			
Operations	X		
Technology development		X	X
Complexity			
Few components or processes			
Moderate number of components or processes	X		
Many components or processes		X	X
Technology maturity	Low	Low	High
Type of innovation			
Regular			X
Niche			
Revolutionary	X	X	
Architectural	X	X	X
Technological competencies	**Laminar**	**Semi**	**Electronics**
Generic engineering skills			
Biological	X		
Chemical	X	X	
Civil			
Computer science		X	
Electrical		X	X
Materials	X	X	
Mechanical	X		X
Physics	X	X	X
Specific engineering skills for clean room			
Filtration	X		
Fluid dynamics	X		
Thermodynamics	X		
Particulate testing	X		
Specific technologies			
Recirculation technologies	X		
Clean room protocols	X		
Particulate control& typing	X		

ability was actually needed due to the adoption of the laminar clean room technology.

Consequently, concentration could be limited to the large number of abilities required by the laminar clean room technology that is currently missing in terms of new managerial capabilities and technical competencies. Semiconductor firms are unaccustomed to *fabrication and assembly* of products as semiconductor processes are material and chemical in nature. However, at that time the industry had not developed a suitable supply base for production equipment. Consequently, many firms were fabricating and assembling at least part of their production equipment. Hence, these skills were already in house to put together the new air-handling systems. The service component associated to the clean room systems is *knowledge embedded*. Contamination sensors were used to identify the point at which the filters for the clean room(s) needed to be changed. While this service component was new to the existing operation, it did not require sophistication and was easily integrated into existing organizational routines. While semiconductor and electronics focused on *technology development*, clean room technology focused on *operations*. As the quantity of clean room systems was small, the inefficiencies in their production by these *technology development* firms were not a problem. As soon as an *operations*-oriented suppler was available, however, users quickly discontinued self-supply. As the complexity associated to laminar flow clean rooms was lower, this difference was not a management challenge for the early users. As this was a new technology, its *technology maturity* was *low*. As electronics firms had been making radios for decades they were accustomed to working with more mature technologies that involved *regular (incremental) innovation*. Having briefly considered the differences in managerial capabilities and their implications, the differences in technological competencies are now addressed.

The *generic engineering skills* overlap quite strongly with the existing skill base in the electronics and semiconductor industries. The exception is that of biology. However, the important biological aspects of clean rooms were of little apparent relevance to these two industries. (Another early adopter of this technology is medical operations—for which the biology of contamination was critical.) What was missing from firms such as RCA and Motorola that pioneered clean room technology use in the manufacture electronics and semiconductors were *specific engineering skills and technologies* associated with clean rooms. Both the *specific engineering skills—filtration, fluid dynamics, thermodynamics,* and *particulate testing*—and the *specific technologies—recirculation technologies, clean room protocols,* and *particulate control and typing*—had to be developed in-house by the adopting firms. At the time of this example 1950s–1960s, foresight and roadmapping techniques were poorly developed so developments and discoveries were more reliant on serendipity, search costs were higher, and the likelihood of mistakes and failure were greater. The utilization of foresight techniques and their integration with the proposed framework allow for greater efficiency and effectiveness through a more rapid and complete identification of needs and challenges.

4.4 Concluding Notes

Changing from current state to future needed state is typically a difficult, but not an insurmountable task. This task can be made much simpler by determining what competencies and capabilities are required at the earliest possible time. In doing so one is given the greatest amount of time to determine the optimal timing and techniques to close the existing gap. The framework provided in this chapter assists by providing a standard, yet flexible template, for listing the needed competencies and capabilities. It is recommended that one follows a four-step process:

1. Fill out framework based on firm current state.
2. Fill out framework based on foresight exercise.
3. Identify gaps that exist—required in future state, but not current state.
4. Determine which gaps should be closed and what steps will be taken to close the gaps that are deemed important.

In doing so one is able to determine and document what is needed, what will be done, and how it will be accomplished. Having such a record not only helps initiate preparation for the future at an early time, but assists in responsively and dynamically changing path as it is determined that the future state appears different from how it was anticipated earlier on and/or that the present state has changed with the passage of time.

Acknowledgement The article was prepared within the framework of the Basic Research Programme at the National Research University Higher School of Economics (HSE) and supported within the framework of the subsidy granted to the HSE by the Government of the Russian Federation for the implementation of the Global Competitiveness Programme.

References

Abernathy WJ, Clark KB (1985) Innovation: mapping the winds of creative destruction. Res Pol 14:3–22
Adler PS (1989) Technology strategy: a guide to the literatures research in technology innovation. Manage Pol 4:25–121
Alvarez SA, Busenitz LW (2001) The entrepreneurship of resource-based theory. J Manage 27 (6):755–775
Barney J (1991) Firm resources and sustained competitive advantage. J Manage 17(1):99–120
Baumgartner P, Wise R (1999) Go downstream: the new profit imperative in manufacturing. Harv Bus Rev 50(5):133–141
Bhave MP (1994) A process model of entrepreneurial venture creation. J Bus Ventur 9(3):223–242
Bitindo D, Frohman A (1981) Linking technological and business planning. Res Manag: 19–23
Bohn RE (1993) Measuring and managing technical knowledge. Sloan Manag Rev 15:61–74
Burns T, Stalker G (1961) The management of innovation. Tavistock, London
Fusfeld AR (1970) The technological progress function: a new technique for forecasting. Technol Forecast 1:301–312
Fusfeld AR (1978) Hot to put technology into corporate planning. Technol Rev 80:63–74

Gomory RE (1989) From the 'ladder of science' to the product development cycle. Harv Bus Rev 67:99–105

Howells J (1997) Rethinking the market-technology relationship for innovation. Res Pol 25 (8):1209–1219

Leiblein MJ, Reuer JJ, Dalsace F (2002) Do make or buy decisions matter? The influence of organizational governance on technological performance. Strat Manag J 23(9):817–833

Leonard-Barton D (1992) Core capabilities and core rigidities: a paradoxin managing new product development. Strat Manag J 13:111–125

Marino KE (1996) Developing a consensuses of firm competence and capabilities. Acad Manag Executive 10(3):40–51

Matzler K, Hinterhuber HH (1998) How to make product development projects more successful by integrating Kano's model of customer satisfaction into quality function deployment. Technovation 18(1):25–38

Michaelis M, Coates JF (1994) Creating integrated performance systems: the business of the future. Technol Anal Strat Manag 2:245–250

Mintzberg H (2007) Tracking strategies. Oxford University Press, New York

Morgan G (2006) Images of organizations. Sage, Thousand Oaks, CA

Munro H, Noori H (1988) Measuring commitment to new manufacturing technology: integrating technological push and market pull concepts. IEEE Trans Eng Manag 35(1):65–70

Park JS (2005) Opportunity recognition and product innovation in entrepreneurial hi-tech start-ups: a new perspective and supporting case study. Technovation 25(7):739–752

Prahalad CK, Hamel G (1990) The core competence of the corporation. Harv Bus Rev 68:79–91

Prahalad CK, Hamel G (1991) Corporate imagination and expeditionary marketing. Harv Bus Rev 69(4):81–92

Schumpeter JA (1934) The theory of economic development. Harvard University Press, Cambridge

Solow R (1957) Technical change and the aggregate production function. Rev Econ Stat 39:312–330

Sullivan LP (1986) Quality function deployment. Quality Progress 19(6):39–50

Thompson CJ (1997) Interpreting consumers: a hermeneutical framework for deriving marketing insights from the texts of consumers' consumption stories. J Market Res 34(4):438–455

Utterback JM (1994) Mastering the dynamics of innovation: how companies can seize opportunities in the face of technological change. Harvard Business School Press, Boston, MA

Veugelers R, Cassiman B (1999) Make and buy in innovation strategies: evidence from Belgian manufacturing firms. Res Pol 28(1):63–80

Von Hippel E (1976) The dominant role of users in the scientific instrument innovation process. Res Pol 5:212–239

von Hippel E (1986) Lead users: a source of novel product concepts. Manag Sci 32(7):791–805

von Hippel E (1988) The sources of innovation. Oxford University Press, New York

Walsh ST, Linton JD (2001) The competence pyramid: a framework for identifying and analyzing firm and industry competence. Technol Anal Strat Manag 13(2):165–177

Walsh ST, Linton JD (2011) The strategy-technology firm fit audit: a guide to opportunity assessment and selection. Technol Forecast Soc Change 78:199–216

Walsh ST, Kirchhoff BA, Newbert S (2002) Differentiating market strategies for disruptive technologies. IEEE Trans Eng Manag 49(4):341–351

Challenges and Opportunities for Corporate Foresight

5

Konstantin Vishnevskiy and Oleg Karasev

5.1 The Emergence of Foresight

The twenty first century is marked by the evolution of new challenges for companies arising from more rapid and drastic changes of socio-economic framework conditions. It turns out that especially companies in knowledge-intensive industries with above average innovation intensity and resource restrictions are under significant pressure of applying reliable instruments for setting priorities. One of the tools to address these challenges is corporate foresight which aims at providing reliable information about potential threats and opportunities which are releant for the companies' activities. Corporate Foresight is therefore understood a special form of foresight which reflects the strategic intentions of companies in different forms and especially takes into account availability of resources and integration of foresight results in company strategies at different levels. Hence corporate foresight is applicable at different stages of the innovation value chain, e.g. the innovation process. Von der Gracht et al. (2010) argue that the main role of foresight is proividing information about potential framework conditions under which corporations operate. This implies that corporate foresight provides a contribution to product development and the innovation pipeline development as one input parameter for innovation. Moreover corporate foresight can contribute to

K. Vishnevskiy (✉)
Institute for Statistical Studies and Economics of Knowledge, National Research University Higher School of Economics, 20 Myasnitskaya Street, 101000 Moscow, Russia
e-mail: kvishnevsky@hse.ru

O. Karasev
Institute for Statistical Studies and Economics of Knowledge, National Research University Higher School of Economics, 20 Myasnitskaya Street, 101000 Moscow, Russia

Faculty of Economics, Lomonosov Moscow State University, Leninskiye Gory, 119991 Moscow, Russia
e-mail: okarasev@hse.ru

© Springer International Publishing Switzerland 2016
L. Gokhberg et al. (eds.), *Deploying Foresight for Policy and Strategy Makers*, Science, Technology and Innovation Studies, DOI 10.1007/978-3-319-25628-3_5

innovation in a sense as being one of the tools for estimating technological progress and technology related feasibility. Accordingly we assume that corporate foresight is also useful for strategy development in different aspects, thus this approach needs to be extended beyond the information delivery function towards a significant contribution to product development and the innovation cycle as a whole. Besides it has a more significant role to play in the complex organization of firms' innovation activities but even beyond.

The term "foresight" ("Look to the Future") appears in literature from the 1960s. These papers provided evidence for the necessity of multivariate and especially multidimensional analysis of the development of future technologies. Similarly work about the role of foresight for management and long-term planning in industry began (Mason 1969; Muther 1969). In the 1970s global foresight, e.g. foresight of scientific development, as well as foresight for a particular country or region was conducted for various purposes, in particular to foresee the potential political and innovation development of countries. However, despite the fact that most of the works offered analysis of foresight techniques for process development, an interest appeared in foresight for the political and social sphere. However the main motivation for using the foresight methods is the technology sphere because in this field the range of possible scenarios, and the factors that may affect these events, is more obvious and accountable (Bronwell 1972; Erickson 1977).

In 1980s foresight at local scale, e.g. first at the regional level and then at the level of individual firms and organizations attracted much attention. This reflects a characteristic feature of the development of foresight techniques—from the global to the local scale (Gokhberg 2016; Sethi 1982; Wee et al. 1989; Martin and Irvine 1984). Later in the early 1990s both theoretical and applied research foresight methodologies received stronger attention, i.e. a broad range of papers on technological foresight, political development foresight, and the foresight of individual organizations were published showing that foresight was actively conducted at both the global and local levels. This is due to the fact that strategic planning was recognized to be effective which required the scientifically founded methods for long-term planning (i.e. foresight and roadmaps) as necessary instruments for successful development (van Dijk 1991; Remmers 1991). The end of the decade was characterized by a sharp increase of the number of publications which positioned foresight as a competitive advantage of one entity over another in the long-term development. Literature focused on the analysis of implementing foresight in business, and in particular, the involvement of senior management in the initiation and creation of foresight. More attention was paid to the analysis of individual components and tool techniques of foresight: SWOT-analysis, scanning technology, extrapolation, etc. (van Wyk 1997; Kuwahara 1999).

In the 2000s scholars were interested in learning what benefits managers can get from the synergies of foresight, mainly studying mechanisms of creating a competitive advantage with foresight. Also the shortcomings of foresight and possible approaches to eliminate them, i.e. situations in which the application of foresight is difficult or impossible received special interest. In addition, researchers were becoming interested in the further refining and developing roadmaps and impact assessment which can be attributed to the development and application of roadmaps. Moreover, there is an ongoing debate on the legality and effectiveness of the integration of scientific methods in the long-term strategic planning (Kappel 2001; Becker 2002; Phaal et al. 2004; Ratcliffe 2006; Daheim and Uerz 2008; Schwarz 2008; Coates et al. 2010; Karasev and Vishnevskiy 2011).

In this chapter we propose a model for conducting corporate foresight and discuss the interfaces with the broader coresight framework. We review existing theoretical approaches to and practical experience of corporate foresight of transnational companies such as Shell, DaimlerChrysler, BASF, Philips, Deutsche Bank etc. The suggested corporate foresight methodology is an outcome of large-scale corporate foresight exercises in the field of long-term future studies for companies (Vishnevskiy et al. 2015a, b; Doroshenko et al. 2011; Khripunova et al. 2014; Kindras et al. 2014; Vishnevskiy and Egorova 2015).

5.2   Corporate Foresight Approaches and Methodologies

Corporate Foresight delivers various different inpouts for company innovation strategies. Therefore a broad range of methodologies is applied which require dedicated organizational approaches for corprate foresight. Ruff (2006) describes three main models of corporate foresight:

1. Companies with in-house or closely affiliated units (*think tank model*) with a stable sizeable team (more than ten persons), diversity of tasks, clear professional identity and continuity (e.g., Shell Planning Group, Deutsche Bank Research, Toyota Gendai, Daimler AG Society and Technology Research Group (STRG)).
2. Integrated functions as part of strategy, innovation or design units with a professional identity as '*futures researchers*' and some (three to ten) team members specialised in this field (e.g., Deutsche Telekom, Philips Design, Henkel).
3. *Mixed models*, dealing temporarily with futures issues, e.g., in cross-functional project teams working on an assigned task, or single outstanding persons representing futures thinking for the company.

The *think tank model* in principle aims at allowing the respective unit to conduct foresight independent from third parties interests, e.g. a comparable high degree of independent blue sky work. However given the fact that these units are frequently integrated in the organizations with the duty to offer services to business units and other corporate entities also think-tank model organized corporate foresight activities aren't free of influence from the external interests and are under pressure to find company internal customers for their initial services. Moreover think tanks are frequently run as a cost centre which makes it challenging for them to justify their activities. This might eventually lead to negative impact on their operations.

The *futures researchers model* is a less independet model which limits the number of staff involved in these activities. Accordingly this model can be considered a flexible approach under which according to the agenda teams can be composed who have necessary competences for different foresight exercises. Although this model ensures a close interaction of foresight staff with actual company activities and challenges there is a threat that staff might be less focused and concetrated at foresight work eventually.

The *mixed models* organization is a compromise between the think-tank and the futures researchers model. However to function effectively this model needs a champion who holds a powerful and influential position in the company.

Table 5.1 shows an overview of corporate foresight in different companies. The fundamental work was done by Kappel (2001) who analysed the effects of roadmapping and its assessment, necessity and estimation of roadmapping quality. In this work roadmaps are both forecasts of what is possible or likely to happen, as well as plans that articulate a course of action. Furthermore his work contains a specification of various kinds of roadmaps and their relation to each other. In addition, research includes case studies that were selected from several large industrial firms. Kappel reveals when the products are complex systems, the price-performance expectations of customers are moving rapidly, or the firm's structure is complex or distributed, the need for roadmaps is high. When a strategic, discontinuous change approaches from the outside, the roadmapping process may not provide early warning.

There is a range of studies that describe methods and approaches of foresight and roadmapping in multinational companies. *Shell* applies scenario-based methodology in order to integrate the foresight activities between its corporate, business and operational levels (Schoemaker and van der Heijden 1992; Vecchiato and Roveda 2010). The core of Shell strategic foresight are global scenarios, which provide a comprehensive analysis of development of the energy industry including different energy sources (oil, gas, renewable). Global scenarios describe changes in politics, economy, society, ecology, technology and demographics at international level. Focused scenarios are much more detailed and based on the global scenarios. The focused scenarios research each business field of the energy industry and each geographic area (USA, Europe, Asia) where the firm works. The project scenarios gather and process more in-depth information on competitors, price, profitability, technical and managerial risk than global and focused scenarios. The time horizon for scenarios is 20 years that are renewed every 3 years taking into account changes in macro trends and their impacts.

Strategic foresight at *BASF* operates with scenarios and has a top-down process that starts at corporate and afterwards regional and business level (Vecchiato and Roveda 2010). Corporate scenarios analyze the global economy and the overall chemical industry for the different regions (EU, USA, Asia) and countries, split into the main sectors and the main business areas of the firm, i.e. chemicals, plastics, performance products, agriculture and nutrition products, oil and gas. Country and business scenarios observe the geographic and business areas of the firm, and its specific innovation and investment. They have a more detailed analysis, which

Table 5.1 Review of corporate foresight studies

Paper	Research aim	Major methods and approaches	Regions	Kind of enterprises
Schoemaker and van der Heijden (1992)	To present a case study demonstrated mechanisms of using scenarios by Shell	• Desk research • Scenarios • Hindsight	The Netherlands	Multinational companies (Shell)
Vecchiato and Roveda (2010)	To present several models of corporate foresight and provide case studies illustrated its process and outcomes	• Scenarios • Roadmaps • Case study	The Netherlands, Germany, Finland	Multinational companies (Shell, BASF, Nokia, Phillips)
Battistella (2014)	To design organization structure of the company to anticipate future trends and detect weak signals	• Scanning • Scenario writing • Assessment • Modeling	Italy	Telecommunication companies
Heger and Rohrbeck (2012)	To create an integrated approach that combines multiple strategic foresight methods in a synergetic way	• Workshops • Interviews • Desk research • Panel discussion	Belgium, France, Germany, The Netherlands, Spain, The UK	Telecommunication companies
Kappel (2001)	To define the practice from other management activities; metrics to evaluate roadmapping performance and impact on the firm; the appropriate circumstances for its application	• Case study • Interviews • Roundtable discussions	—	Large industrial firms
Lee et al. (2011)	To analyse the factors influencing the utilization of TRMs (survey of 186 different R&D units)	• R&D Survey • Questionnaire • Multiple regression analysis	Republic of Korea	Stock market-listed companies

(continued)

Table 5.1 (continued)

Paper	Research aim	Major methods and approaches	Regions	Kind of enterprises
Rohrbeck et al. (2007)	To identify links between strategic foresight with other functions in companies	• Roadmapping • Scenarios • Quality scenarios • Delphi	Germany	Multinational company (Deutsche Telekom)
Rohrbeck (2008)	To develop capability model for strategic foresight practices	• Literature review • Interviews	Germany, Netherlands, UK, Portugal, Austria	Multinational companies
Rohrbeck and Gemünden (2011)	To analyze the ways of integration of foresight and innovation strategies of companies	• Literature review • Interviews • Workshops	Germany, Sweden, Spain, Netherlands, USA, UK, Portugal, Austria	Multinational companies
Rohrbeck and Schwarz (2013)	The reveal potential and empirically observable value creation of strategic foresight activities in firms	• Literature review • Interviews	European countries	Multinational companies (annual revenue > €100 million)
Ruff (2006)	To observe the initiation of corporate foresight, major working areas and benefits for the company	• Literature review • International experts interviews • Workshops • Desk research • Multi-level scenario approach	China	Multinational automotive company
Thom (2010)	To discuss of possible approaches of value measurement in corporate foresight	• Expert group • Assessment • Workshops	Germany	Multinational company (Deutsche Telekom)

Sources: compiled by National Research University Higher School of Economics considers a larger range of objectives such as national regulations or exchange rates, and market issues. For global scenarios the time frame is usually 10–15 years.

The main task of *Philips* foresight activities is to identify new trends in society, customers' needs and technology (Groenveld 1997). For this purpose the Philips Group has an independent unit (Philips Design) which delivers design services for the different businesses of the corporation and for clients outside the firm. Philips Design investigates the three axes of 'society', 'culture' and 'people'. Moreover there are some extra actors involved in foresight studies in Philips, namley Philips Research, the R&D corporate unit of the group. The roadmaps produced during corporate foresight studies include new product development, prioritization of R&D programmes and exploitation of synergies among technologies. As a result it brings together social researchers from Philips Design, technologists from Philips Research and business managers from all product divisions. Strategic foresight activities usually cover a time horizon of 10 years. The identification of emerging trends is conducted every year to update company's foresight results.

Daimler AG (STRG) was one of the first foresight groups to be established within a company that has operated since 1979 (Ruff 2006). The research team includes about 40 interdisciplinary researchers. The analyses are conducted on macro-, meso- and micro- levels with a time horizon spanning medium- and long-term periods (5–15 years). To realise successfully international projects STRG employs an international network of partners in Europe, the USA and Asia. STRG integrates several methodological building blocks including scenarios; product assessment; conceptualization of future consumer needs and desires; multimedia display of business environment and social trends; strategic options prioritization; expert panel assembly on short notice; innovation processes design, organization systems design; development of think tools to organize new forms of communication and decision-making. In his study Ruff (2006) focuses on the practice of corporate foresight within a multinational automotive company. The target of the study is to observe the initiation of corporate foresight, major working areas and benefits for the company. Foresight activities in STRG allow identifying opportunities and risks and support the generation, development and evaluation of innovative ideas. The results of scenario analysis are recommendations that include statements about the business environment and organization's strengths and weaknesses. The opportunities and risks facing the company in different scenarios are translated into an action agenda.

Deutsche Bank Research (DBR) is the independent unit of Deutsche Bank Group that delivers foresight for the company's decision-makers, clients and the public. DBR foresight activities primarily aim at enhancing the competitiveness of Deutsche Bank. Moreover foresight applies different time perspectives, thus short-, medium- and long-term trends are covered. In DBR a synthesis of different methods is developed. It contains combination of mathematical, statistical methods and econometric modeling with scenario analysis, horizon scanning, trend analysis, driver analysis, and visioning of paths as well as qualitative methods from social sciences. The long-term projection of future growth paths of national economies are observed in the DBR "trend map" that extends over 20–40 years and show "MacroTrends".

Foresight activities at *Deutsche Telekom* allow cutting uncertainty and the exploration of new business fields involving three units for corporate foresight. The first one is a competitor foresight with its major tool named the Product & Service Radar. An international network of scouts is used for the identification of developments and trends in the market. The second, technology foresight is presented by the Technology Radar. It gives an opportunity to identify expert groups, assess and analyze technological developments and trends based on technology researches (Rohrbeck et al. 2007). The third, customer foresight helps to identify, assess and anticipate of consumer needs, lifestyle and socio-cultural trends. Deutsche Telekom Laboratories (T-Labs) are part of the central 'Innovation, Research, and Development' unit of Deutsche Telekom AG. The T-Labs research and develop new information and communication technologies and services by combining scenario analysis and roadmapping. The experience of Deutsche Telekom shows a successful strategic foresight activity that is based on a deep understanding of the need of the decision maker, involvement of many internal and external partners and use of a balanced mix of qualitative and quantitative methods.

Rohrbeck (2008) observes the employment of strategic foresight by corporations (e.g. Deutsche Telekom, British Telekom, Telekom Austria, Philips, Osram, Continental, Thyssen Krupp Automotive, Vattenfall Europe, EDP). Based on the model of Day and Schoemaker (2005) the author has developed the framework for a strategic foresight capability model. It stands to mention that this study uses companies that are different from each other in terms of industry, position in the value chain, and from their primary business driver, which could be either technology driven of market driven. Afterwards Rohrbeck and Gemünden (2010) have extended this original research and observe the ability of corporate foresight to increase the innovation capacity of a firm. They reveal three roles that corporate foresight should play to maximize the innovation capacity of a firm: (1) the strategist role, which explores new business fields; (2) the initiator role, which increases the number of innovation concepts and ideas; and (3) the opponent role, which challenges innovation projects to increase the quality of their output.

Thom (2010) considers applying corporate foresight (consumer foresight, competitor foresight, technology foresight) using the example of Deutsche Telekom. according to his study value is created, when insights of corporate foresight activities are turned into action and output, such as enhanced reaction to opportunities and threats and reduced uncertainty for better decision-making is achieved. The paper observes that the value contribution is independent, except of two aspects: corporate culture and organizational structure. A company with no appreciation for ideas and no openness will hardly be able to see a value in ideas coming from corporate foresight. Where technology foresight was directly linked to follow-up activities, it was possible to directly measure the impact or at least to account for an impact. Later Rohrbeck and Schwarz (2013) use survey data from 77 large multinational firms (annual revenue of more than €100million) to assess how much value is generated from formalized strategic foresight practices in these firms. As a result it was noted that possible ways to capture value are (1) an enhanced capacity to perceive change, (2) an enhanced capacity to interpret and respond to change, (3) influencing other actors, (4) an enhanced capacity for organizational learning.

Lee et al (2011) investigate the factors influencing the utilization of technology roadmaps. For this study they analysed 186 different R&D units of stock market-listed companies in the Republic of Korea. The research exhibits that the utilization of technology roadmaps is most significantly influenced by appropriate software, an effective roadmap process, organizational support, and the map's alignment with company objectives.

Other papers analyze organization of corporate foresight in telecommunication companies, for instance Battistella (2014) or Heger and Rohrbeck (2012). Battistella's work (2014) aimed at identification of links between value chains and the typology of foresight studies on the case of Italian telecommunication companies. As a result the study reveals that the manufacturers are more focused on the technological foresight, while the operators in the competitive and consumer foresight studies; while it does not influence the corporate foresight focus on strategy or innovation and R&D. Heger and Rohrbeck (2012) present an integrated study that combines multiple strategic foresight methods in a synergetic way. The authors involved telecommunication companies from Belgium, France, Germany, the Netherlands, Spain, and the UK in their study. It gave them an opportunity to elaborate an integrated foresight methodology that combines qualitative and quantitative approaches and can be used to make companies more reliable and effective.

İn summary it can be concluded that most of the existing approaches are disconnected from the actual strategy development which is a major weakness. Therefore we postulate that corporate foresight should follow an integrated approach.

5.3   Integrated Approach to Corporate Foresight

Market perspectives of future-oriented technologies can hardly be estimated by methods of traditional quantitative forecasts based on previously observable data. Irrelevances of conventional methods together with insufficiencies of generally accessible data themselves hinder the assessment of emergent markets. New heuristics based on wide-ranged expert methods are up to the challenge. Therefore we suggest using a multi-methodology approach for corporate foresight: integration of desk research, expert procedures and scenario writing. The suggested methodology includes two main parts: priority setting and integrated roadmapping (Fig. 5.1).

Priority setting reveals the main trends & challenges of economic, scientific and social development. This is complemented by the consideration of different scenarios of scientific-technological development, points of growth (emerging markets, prospective directions of development, innovation products and technologies) and the creation of a priorities system.

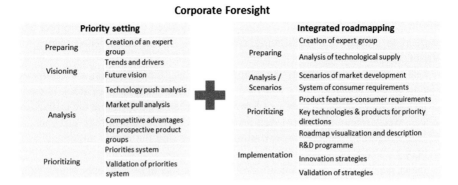

Fig. 5.1 Integrated methodology of corporate foresight. Source: National Research University Higher School of Economics

Integrated roadmapping elaborates an action plan for strategic priorities realization elaborated, the most prospective directions of the company's development in the long-term are highlighted and the "corridor" for the formation of projects prepared. The integrated roadmap incorporates the technology roadmap which interrelates with the most prospective new products, technologies and R&D into business maps that depicts alternative market patterns for these products and technologies. A wide range of different long-term future analysis methods is employed for priority setting and integrated roadmapping (Table 5.2).

Using the multi-faceted methodology of corporate foresight helps to generalize the expert community views on innovative development of the company and integrate qualitative and quantitative analysis. It allows identifying new products, technologies and R&D priorities which are utmost important in achieving the goals set, and to appraise technological forks economically as well as to compare alternative paths of company's future development. This also allows that corporate foresight out comes can be utilized by both internal and external users (Fig. 5.2). Corporate foresight can help to identify and coordinate company priorities, to make scientifically-grounded investment decisions, to identify areas of demand stimulation and to make project expertise. In total it leads to investment risks decrease. This instrument is also very useful in company's interaction with external economic agents because it helps to identify possibilities of effective cooperation with development institutions, technological platforms, clusters and position itself in domestic and international market.

Corporate foresight has promising practical application potential for adjusting companies' priorities, project expertise and appraisal, supporting investment decision-making. It can help to identify the most prospective innovation products as it provides both technological and commercial validation of multiple-choice alternative chains «R&D—technology—product—market». Comparative studies of these chains generate concrete strategic options including technology/product specifications; comparative advantages over analogues; benefit-giving properties; future-oriented fields of application; expected market demand.

Table 5.2 Methods used in corporate foresight

Methods	Corporate foresight	
	Priority setting	Integrated roadmapping
Delphi and survey	X	
Workshop	X	X
Key technologies	X	
Literature review	X	X
Bibliometrics and patent analysis	X	X
Benchmarking	X	X
Scanning	X	
Interviews	X	X
Expert panels	X	X
Wild cards and weak signals		X
Backcasting	X	X
SWOT-analysis	X	X
Cross-impact analysis	X	
Brainstorming	X	X
Stakeholders analysis		X
Scenario workshops		X

Source: National Research University Higher School of Economics

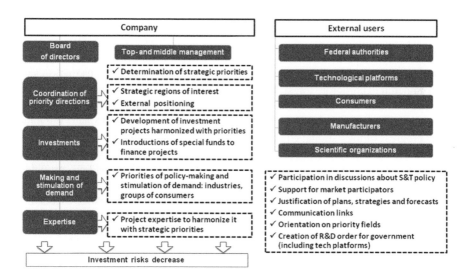

Fig. 5.2 Users of corporate foresight. Source: National Research University Higher School of Economics

Using corporate foresight companies are provided scenarios of research field development; estimation of the commercial perspectives of innovation products and technologies and creation of innovation strategies for customers. Its application

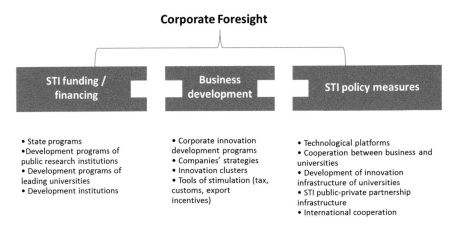

Fig. 5.3 Integration of corporate foresight and other STI activities. Source: National Research University Higher School of Economics

includes programmes for R&D, project and programme implementation and commercialization strategies of future-oriented technologies and products.

The main intention of corporate foresight is contribution to innovation strategies (Karasev et al. 2010, 2013). Thus multifaceted character of the integrated methodology (applicable both for internal and external users) gives an opportunity to implement results of corporate foresight into the national STI system and harmonise company's innovation strategies with other STI activities (Fig. 5.3).

Obviously corporate foresight plays the crucial role in bridging STI funding, business development and STI policy measures which means that information and data gathered in corporate foresight are overlapping with the other three dimensions and are consequently included in corporate foresight. At the same time corporate foresight reflects national STI policy and respective efforts which indicate the long term progress of the national competence base. The latter is important for corporate innovation strategy because it provides firsthand information of companies about potential centres of competence/excellence. Moreover STI funding provides opportunities for companies to leverage own activities with special STI support programmes and taking advantage of the national STI infrastructure, especially universities and public research institutions. İn this regard corporate foresight helps to set a system of priorities considering the national competence base expressed in STI policy initiatives and funding. This is due to the fact that national priorities reflect the expected future availability of skills and competences needed for business development. In a broader sense this change can be viewed as a reflection of an extended open innovation paradigm. Company's business strategies help to realize its cooperation potential which is offered by STI policy measures like technology platforms, public-private partnership, international collaboration etc. Thus, corporate foresight represents an integrated instrument that enhance different shapes and styles of cooperation which are characteristic for open innovation.

5.4 Conclusion

To sum up companies can use corporate foresight for setting strategic objectives and creating roadmaps for their achievement. The systemic approach of strategic decision-making and planning at corporate level embraces the study of a wide portfolio of external and internal factors and trends that have an important impact on the formation of the priorities system of the development and selection of promising innovative projects.

The employment of corporate foresight allows achieving following results:

- a priority system for the medium and long-term periods that takes into account company's competitive advantages in R&D and production;
- strategic objectives of technological development, tailored to technological competition factors, development of demand, the current state of and prospects of R&D and production base;
- target characteristics of innovative products that can (or should) be developed and introduced to the market in the medium and long-term periods;
- the most promising innovative strategy aimed at developing the existing competitive advantages or catching up with the best foreign achievements, including:

  - promising strategies for research and development;
  - promising strategy for the implementation and adaptation of innovative products;

- roadmaps for implementation of the most promising innovation strategies.

The potential of using corporate foresight lies not only in the information generated, processed and visualized but moreover in achieving an overarching impact on corporations' strategic orientation. Corporate foresight in the firm can be considered as a catalysts of shift toward 'future oriented organizational culture' of the firm which is usually supportive to innovation. Thus the cultural function of corporate foresight is one step towards raising awareness for future uncertainties.

Moreover integrated roadmaps as a final part of corporate foresight if applied and updated systematically show significant potential to improve the validity and solidity of strategies by bringing the diverging technology analysis and market projection views together.

Integrated methodology of corporate foresight could help companies to array strategic priorities for their research and technology policy in the field of new technologies in view of scenario-related international, national and industrial technological and market developments.

Acknowledgement The chapter was prepared within the framework of the Basic Research Programme at the National Research University Higher School of Economics (HSE) and supported within the framework of the subsidy granted to the HSE by the Government of the Russian Federation for the implementation of the Global Competitiveness Programme.

References

Battistella C (2014) The organisation of corporate foresight: a multiple case study in the telecommunication industry. Technol Forecast Soc Change 87:60–79

Becker P (2002) Corporate foresight in Europe: a first overview. Working paper. European Commission, Brussels

Bronwell AB (1972) Technological forecasting in the formulation of national policy. Ind Market Manag 1(4):431–439

Coates J, Durance P, Godet M (2010) Strategic foresight issue: introduction. Technol Forecast Social Change 77(9):1423–1425

Daheim C, Uerz G (2008) Corporate foresight in Europe: from trend based logics to open foresight. Technol Anal Strateg Manag 20(3):321–336

Day GS, Schoemaker PJH (2005) Scanning the periphery. Harv Bus Rev 83(11):135–148

Doroshenko M, Vishnevskiy K, Karasev O (2011) Foresight and roadmapping backgrounding public and private innovation strategies: Approach of the HSE. In: Theory building in foresight and futures studies, Book of Abstracts. Yeditepe International Research Conference on Foresight and Futures. p. 64

Erickson J (1977) The Soviet Union, the future, and futures research. Futures 9(4):335–339

Gokhberg L (ed) (2016) Russia 2030: science and technology foresight. Ministry of Education and Science of the Russian Federation, National Research University Higher School of Economics, Moscow

Groenveld P (1997) Roadmapping integrates business and technology. Res Technol Manag 40 (5):48–55

Heger T, Rohrbeck R (2012) Strategic foresight for collaborative exploration of new business fields. Technol Forecast Social Change 79(5):819–831

Kappel TA (2001) Perspectives on roadmaps: how organizations talk about the future. J Product Innov Manag 18(1):39–50

Karasev O, Vishnevskiy K (2011) Development of innovation strategies for government and corporations on the basis of roadmaps: Experience of the HSE, Brief of Future-oriented technology analysis conference, IPTS

Karasev O, Vishnevskiy K (2013) Toolkit for integrated roadmaps: employing nanotechnologies in water and wastewater treatment. In: Meissner D, Gokhberg L, Sokolov A (eds) Science, technology and innovation policy for the future–potentials and limits of foresight studies. Springer, Heidelberg, New York, Dordrecht, London, pp 137–159

Karasev O, Vishnevskiy K (2010) Identifying the future of new materials with the use of foresight methods. Foresight-Russia 4(2):58–67 (in Russian)

Kindras A, Meissner D, Vishnevskiy K, Cervantes M (2014) Regional foresight for bridging national science, technology and innovation with company innovation: experiences from Russia. Working papers by NRU Higher School of Economics. Series WP BRP "Science, Technology and Innovation". No. 29/STI/2014.

Khripunova A, Vishnevskiy K, Karasev O, Meissner D (2014) Corporate foresight for corporate functions: impacts from purchasing functions. Strat Change 23(3–4):147–160

Kuwahara T (1999) Technology forecasting activities in Japan. Technol Forecast Social Change 60(1):5–14

Lee JH, Phaal R, Lee C (2011) An empirical analysis of the determinants of technology roadmap utilization. R&D Manag 41(5):485–508

Martin BR, Irvine J (1984) Foresight in science: picking the winners. Pinter, London

Mason HR (1969) Developing a planning organization: a logical sequence of phases. Bus Horiz 12 (4):61–69

Muther R (1969) Long range planning of industrial facilities. Long Range Plann 2(2):58–60

Phaal R, Farrukh CJP, Probert DR (2004) Technology roadmapping—a planning framework for evolution and revolution. Technol Forecast Social Change 71(1–2):5–26

Ratcliffe JS (2006) Challenges for corporate foresight: towards strategic prospective through scenario thinking. Foresight 8(1):39–54

Remmers HL (1991) If foresight were as clear as hindsight. . . Managing currency fluctuations. Eur Manag J 9(3):247–254

Rohrbeck, R. (2008). Towards a best-practice framework for strategic foresight: Building theory from case studies in multinational companies. In: IAMOT conference proceedings

Rohrbeck R, Gemünden HG (2011) Corporate foresight: its three roles in enhancing the innovation capacity of a firm. Technol Forecast Social Change 78(2):231–243

Rohrbeck R, Schwarz JO (2013) The value contribution of strategic foresight: insights from an empirical study of large European companies. Technol Forecast Social Change 80 (8):1593–1606

Rohrbeck R, Arnold HM, Heuer J (2007) Strategic foresight—a case study on the Deutsche Telekom Laboratories. ISPIM-Asia conference, New Delhi, India, 9th–12th January 2007

Ruff F (2006) Corporate foresight: integrating the future business environment into innovation and strategy. Int J Technol Manag 34(3/4):278–295

Schoemaker PJ, van der Heijden CA (1992) Integrating scenarios into strategic planning at Royal Dutch/Shell. Plann Rev 20(3):41–46

Schwarz O (2008) Assessing the future of futures studies in management. Futures 40(3):237–246

Sethi NK (1982) Strategic planning system for multinational companies. Long Range Plann 15 (3):80–89. doi:10.1016/0024-6301(82)90029-2

Thom N (2010) Measuring the value contribution of corporate foresight. In: 3rd ISPIM Innovation Symposium. Quebec City, Canada

van Dijk JWA (1991) Foresight studies. Technol Forecast Social Change 40(3):223–234. doi:10. 1016/0040-1625(91)90053-I

van Wyk RJ (1997) Strategic technology scanning. Technol Forecast Social Change 55(1):21–38. doi:10.1016/S0040-1625(97)83077-6

Vecchiato R, Roveda C (2010) Foresight in corporate organisations. Technol Anal Strat Manag 22 (1):99–112

Vishnevskiy K, Egorova O (2015) Strategic foresight for SMEs: choice of the most relevant methods. In: The XXVI ISPIM Conference—Shaping the Frontiers of Innovation Management, Budapest, Hungary, 14th–17th June 2015

Vishnevskiy K, Karasev O, Meissner D (2015a) Integrated roadmaps and corporate foresight as tools of innovation management: the case of Russian companies. Technol Forecast Social Change 90:433–443

Vishnevskiy K, Meissner D, Karasev O (2015b) Strategic foresight: state-of-the-art and prospects for Russian corporations. Foresight 17(5):460–474

von der Gracht HA, Vennemann CR, Darkow IL (2010) Corporate foresight and innovation management: a portfolio-approach in evaluating organizational development. Futures 42 (4):380–393

Wee CH, Farley JU, Seok KL (1989) Corporate planning takes off in Singapore. Long Range Plann 22(2):78–90. doi:10.1016/0024-6301(89)90126-X

Technological Evolution and Transhumanism

6

José Cordeiro

The famous astronomer and astrobiologist Carl Sagan popularized the concept of a Cosmic Calendar about three decades ago. In his 1977 book, *The Dragons of Eden: Speculations on the Evolution of Human Intelligence*, Sagan wrote a timeline for the universe, starting with the Big Bang about 15 billion years ago. Today, we think that it all started about 13.7 billion years back, and we keep updating and improving our knowledge of life, the universe and everything. In his Cosmic Calendar, with each month representing slightly over one billion years, Sagan dated the major events during the first 11 months of the cosmic year (see Table 6.1).

Interestingly enough, most of what we study in biological evolution happened in the last month. In fact, Sagan wrote that the first worms appeared on December 16, the invertebrates began to flourish on the 17th, the trilobites boomed on the 18th, the first fish and vertebrates appeared on the 19th, the plants colonized the land on the 20th, the animals colonized the land on the 21st, the first amphibians and first winged insects appeared on the 22nd, the first trees and first reptiles evolved on the 23rd, the first dinosaurs appeared on the 24th, the first mammals evolved on the 26th, the first birds emerged on the 27th, the dinosaurs became extinct on the 28th, the first primates appeared on the 29th and the frontal lobes evolved in the brains of primates and the first hominids appeared on the 30th. Basically, humans are just the new kids in the block, and only evolved late at night on the last day of this Cosmic Calendar (see Table 6.2).

The previous Cosmic Calendar is an excellent way to visualize the acceleration of change and the continuous evolution of the universe. Other authors have

J. Cordeiro, Ph.D. (✉)
Venezuela Node Department, The Millennium Project, Avenida Rómulo Gallegos, Residencias Italia #37, Caracas, Venezuela

Singularity University, NASA Research Park, Moffett Field, CA, USA

Venezuela Chapter, World Future Society, Caracas, Venezuela
e-mail: jose@millennium-project.org

© Springer International Publishing Switzerland 2016
L. Gokhberg et al. (eds.), *Deploying Foresight for Policy and Strategy Makers*,
Science, Technology and Innovation Studies, DOI 10.1007/978-3-319-25628-3_6

Table 6.1 Cosmic calendar: January to November

Big Bang	January 1
Origin of Milky Way Galaxy	May 1
Origin of the solar system	September 9
Formation of the Earth	September 14
Origin of life on Earth	~ September 25
Formation of the oldest rocks known on Earth	October 2
Date of oldest fossils (bacteria and blue-green algae)	October 9
Invention of sex (by microorganisms)	~ November 1
Oldest fossil photosynthetic plants	November 12
Eukaryotes (first cells with nuclei) flourish	November 15

Source: Cordeiro based on Sagan (1977)

developed similar ideas to try to show the rise of complexity in nature. For example, in 2005, astrophysicist Eric Chaisson published *Epic of Evolution: Seven Ages of the Cosmos*, where he describes the formation of the universe through the development of seven ages: matter, galaxies, stars, heavy elements, planets, life, complex life, and society. Chaisson presents a valuable survey of these fields and shows how combinations of simpler systems transform into more complex systems, and he thus gives a glimpse of what the future might bring.

Both Sagan and Chaisson have written excellent overviews about evolution, from its cosmic beginnings to the recent emergence of humans and technology. However, a more futuristic look is given by engineer and inventor Ray Kurzweil in his 2005 book: *The Singularity is Near: When Humans Transcend Biology*. Kurzweil wrote about six epochs with increasing complexity and accumulated information processing (see Table 6.3).

According to Kurzweil, humanity is entering Epoch 5 with an accelerating rate of change. The major event of this merger of technology and human intelligence will be the emergence of a "technological singularity". Kurzweil believes that within a quarter century, non-biological intelligence will match the range and subtlety of human intelligence. It will then soar past it because of the continuing acceleration of information-based technologies, as well as the ability of machines to instantly share their knowledge. Eventually, intelligent nanorobots will be deeply integrated in our bodies, our brains, and our environment, overcoming pollution and poverty, providing vastly extended longevity, full-immersion virtual reality incorporating all of the senses, and vastly enhanced human intelligence. The result will be an intimate merger between the technology-creating species and the technological evolutionary process it spawned.

Computer scientist and science fiction writer Vernor Vinge first discussed this idea of a technological singularity in a now classic 1993 paper, where he predicted:

> Within thirty years, we will have the technological means to create superhuman intelligence. Shortly after, the human era will be ended.

Table 6.2 Cosmic calendar: December 31

Origin of *Proconsul* and *Ramapithecus*, probable ancestors of apes and men	~1:30 p.m.
First humans	~10:30 p.m.
Widespread use of stone tools	11:00 p.m.
Domestication of fire by Peking man	11:46 p.m.
Beginning of most recent glacial period	11:56 p.m.
Seafarers settle Australia	11:58 p.m.
Extensive cave painting in Europe	11:59 p.m.
Invention of agriculture	11:59:20 p.m.
Neolithic civilization; first cities	11:59:35 p.m.
First dynasties in Sumer, Ebla and Egypt; development of astronomy	11:59:50 p.m.
Invention of the alphabet; Akkadian Empire	11:59:51 p.m.
Hammurabi legal codes in Babylon; Middle Kingdom in Egypt	11:59:52 p.m.
Bronze metallurgy; Mycenaean culture; Trojan War; Olmec culture; invention of the compass	11:59:53 p.m.
Iron metallurgy; First Assyrian Empire; Kingdom of Israel; founding of Carthage by Phoenicia	11:59:54 p.m.
Asokan India; Ch'in Dynasty China; Periclean Athens; birth of Buddha	11:59:55 p.m.
Euclidean geometry; Archimedean physics; Ptolemaic astronomy; Roman Empire; birth of Christ	11:59:56 p.m.
Zero and decimals invented in Indian arithmetic; Rome falls; Moslem conquests	11:59:57 p.m.
Mayan civilization; Sung Dynasty China; Byzantine empire; Mongol invasion; Crusades	11:59:58 p.m.
Renaissance in Europe; voyages of discovery from Europe and from Ming Dynasty China; emergence of the experimental method in science	11:59:59 p.m.
Widespread development of science and technology; emergence of global culture; acquisition of the means of self-destruction of the human species; first steps in spacecraft planetary exploration and the search of extraterrestrial intelligence	Now: The first second of New Year's Day

Source: Cordeiro based on Sagan (1977)

Other authors talk about such technological singularity as the moment in time when artificial intelligence will overtake human intelligence. Kurzweil has also proposed the *Law of Accelerating Returns*, as a generalization of Moore's law to describe an exponential growth of technological progress. Moore's law deals with an exponential growth pattern in the complexity of integrated semiconductor circuits (see Fig. 6.1).

Kurzweil extends Moore's law to include technologies from far before the integrated circuit to future forms of computation. Whenever a technology approaches some kind of a barrier, he writes, a new technology will be invented to allow us to cross that barrier. He predicts that such paradigm shifts will become

Table 6.3 The six epochs of the universe according to Kurzweil

Epoch 1	Physics and chemistry (information in atomic structures)
Epoch 2	Biology (information in DNA)
Epoch 3	Brains (information in neural patterns)
Epoch 4	Technology (information in hardware and software designs)
Epoch 5	Merger of technology and human intelligence (the methods of biology, including human intelligence, are integrated into the exponentially expanding human technology base)
Epoch 6	The universe wakes up (patterns of matter and energy in the universe become saturated with intelligent processes and knowledge)

Source: Cordeiro based on Kurzweil (2005)

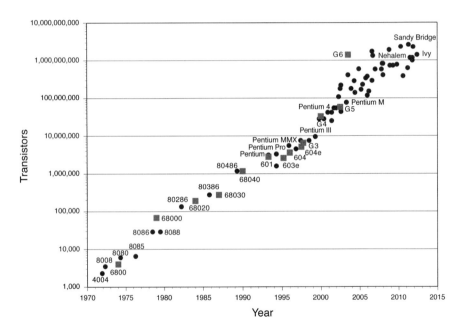

Fig. 6.1 Moore's Law. Source: Based on Intel

increasingly common, leading to "technological change so rapid and profound it represents a rupture in the fabric of human history." He believes the *Law of Accelerating Returns* implies that a technological singularity will occur around 2045:

> An analysis of the history of technology shows that technological change is exponential, contrary to the common-sense 'intuitive linear' view. So we won't experience 100 years of progress in the 21st century—it will be more like 20,000 years of progress (at today's rate). The 'returns,' such as chip speed and cost-effectiveness, also increase exponentially. There's even exponential growth in the rate of exponential growth. Within a few decades, machine intelligence will surpass human intelligence, leading to the Singularity—

technological change so rapid and profound it represents a rupture in the fabric of human history. The implications include the merger of biological and non-biological intelligence, immortal software-based humans, and ultra-high levels of intelligence that expand outward in the universe at the speed of light.

6.1 Technological Convergence

Futurists today have diverging views about the singularity: some see it as a very likely scenario, while others believe that it is more probable that there will never be any very sudden and dramatic changes due to progress in artificial intelligence. However, most futurists and scientists agree that there is an increasing rate of technological change. In fact, the rapid emergence of new technologies has generated scientific developments never dreamed of before.

The expression "emerging technologies" is used to cover such new and potentially powerful technologies as genetic engineering, artificial intelligence, and nanotechnology. Although the exact denotation of the expression is vague, various writers have identified clusters of such technologies that they consider critical to humanity's future. These proposed technology clusters are typically abbreviated by such combinations of letters as NBIC, which stands for Nanotechnology, Biotechnology, Information technology and Cognitive science. Various other acronyms have been offered for essentially the same concept, such as GNR (Genetics, Nanotechnology and Robotics) used by Kurzweil, while others prefer NRG because it sounds similar to "energy." Journalist Joel Garreau in *Radical Evolution* uses GRIN, for Genetic, Robotic, Information, and Nano processes, while author Douglas Mulhall in *Our Molecular Future* uses GRAIN, for Genetics, Robotics, Artificial Intelligence, and Nanotechnology. Another acronym is BANG for Bits, Atoms, Neurons, and Genes.

The first NBIC Conference for Improving Human Performance was organized in 2003 by the NSF (National Science Foundation) and the DOC (Department of Commerce). Since then, there have been many similar gatherings, in the USA and around the world. The European Union has been working on its own strategy towards converging technologies, and so have been other countries in Asia, starting with Japan.

The idea of technological convergence is based on the merger of different scientific disciplines thanks to the acceleration of change on all NBIC fields. Nanotechnology deals with atoms and molecules, biotechnology with genes and cells, infotechnology with bits and bytes, and cognitive science with neurons and brains. These four fields are converging thanks to the larger and faster information processing of ever more powerful computers (see Fig. 6.2).

Experts from the four NBIC fields agree about the incredible potential of technological evolution finally overtaking and directing biological evolution. Bill Gates of Microsoft has stated that:

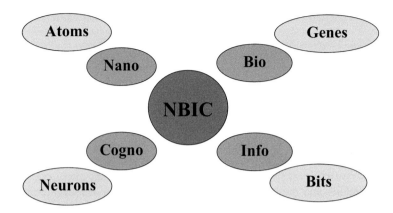

Fig. 6.2 Technological Convergence NBIC. Source: Cordeiro based on Roco and Bainbridge (2003)

> I expect to see breathtaking advances in medicine over the next two decades, and biotechnology researchers and companies will be at the centre of that progress. I'm a big believer in information technology... but it is hard to argue that the emerging medical revolution, spearheaded by the biotechnology industry, is any less important to the future of humankind. It, too, will empower people and raise the standard of living.

Larry Ellison of Oracle, Gates' rival in the software industry, agrees: "If I were 21 years old, I probably wouldn't go into computing. The computing industry is about to become boring". He explains that: "I would go into genetic engineering." Biologist Craig Venter has said that he spent 10 years reading the human genome, and now he is planning to write new genomes. He wants to create completely new forms of life, from scratch. Indeed, Venter and his team created in 2010 the first synthetic bacterium, and they called it *Synthia*. Scientist Gregory Stock also believes that cloning, even though a fundamental step in biotechnology, is just too simple and unexciting: "why copy old life forms when we can now create new ones?"

Biological evolution allowed the appearance of human beings, and many other species, through millions of years of natural selection based on trials and errors. Now we can control biological evolution, direct it and go beyond it. In fact, why stop evolution with carbon-based life forms? Why not move into silicon-based life, among many other possibilities? Robotics and artificial intelligence will allow us to do just that.

Scientist Marvin Minsky, one of the fathers of artificial intelligence at MIT, wrote in 1994 his very famous article "Will robots inherit the Earth?" in *Scientific American*, where he concludes: "Yes, but they will be our children. We owe our minds to the deaths and lives of all the creatures that were ever engaged in the struggle called Evolution. Our job is to see that all this work shall not end up in meaningless waste." Robotics expert Hans Moravec has written two books about robots and our (their) future: *Mind Children* in 1988 and *Robot* in 1998. Moravec argues that robots will be our rightful descendants and he explains several ways to "upload" a mind into a robot. In England, cybernetics professor Kevin Warwick has been implanting his own body with several microchip devices and published in 2003 a book explaining his experiments: *I, Cyborg*. Warwick is a cybernetics pioneer who claims that: "I was born human. But this was an accident of fate—a condition merely of time and place. I believe it's something we have the power to change... The future is out there; I am eager to see what it holds. I want to do something with my life: I want to be a cyborg."

As these authors and thinkers suggest, we need to start preparing ourselves for the coming NBIC realities of technological convergence, including robotics and artificial intelligence. Thanks to technological evolution, humans will transcend our biological limitations to become transhumans and eventually posthumans. To ease this transition into a posthuman condition, we must ready ourselves for the distinct possibility that the Earth, and other planets, will be inherited by not just one but several forms of highly intelligent and sentient life forms. Thus, the philosophy of humanism is not enough for a world, and a universe, where future life forms will continue evolving.

6.2   From Humanism to Transhumanism

A new philosophy has been proposed to continue the ideas of humanism in a world where science and technology are the major drivers of change. Julian Huxley, the English evolutionary biologist and humanist that became the first director-general of UNESCO and founder of the World Wildlife Fund, wrote that:

> The human species can, if it wishes, transcend itself—not just sporadically, an individual here in one way, an individual there in another way, but in its entirety, as humanity. We need a name for this new belief. Perhaps transhumanism will serve: man remaining man, but transcending himself, by realizing new possibilities of and for his human nature.
>
> "I believe in transhumanism": once there are enough people who can truly say that, the human species will be on the threshold of a new kind of existence, as different from ours as ours is from that of Pekin man. It will at last be consciously fulfilling its real destiny.

Huxley originally published those words in his essay *Religion Without Revelation* (1927), which was later reprinted in his book *New Bottles for New Wine* (1957). Other scientists and philosophers discussed similar ideas in the first half of the twentieth century, and these ideas slowly helped to create new philosophical movements considering nature and humanity in a continuous state of flux and evolution. English scientist John Burdon Sanderson (JBS) Haldane and French philosopher Pierre Teilhard de Chardin helped to identify new trends in the future evolution of humanity. Thanks to them and many others, the philosophy of transhumanism has greatly advanced since Huxley first used that word. The philosophy of Extropy (see Appendix 1) and Transhumanism (see Appendix 2) explore the boundless opportunities for future generations, while we approach a possible technological singularity.

"Humans" can no longer be regarded as a stable category let alone one which occupies a privileged position in relation to all that is subsumed under the category of the non-human. On the contrary, humans must be understood as a tenuous entity which is related to the animal, the "natural" and indeed other humans as well. Humans are at a crossroads like other natural species that are reclassified in the face of new relational dynamics and shifting epistemological paradigms. Moreover, such dynamics and interpolation serve to reveal the boundaries of humans as a corporal, cognitive, and agency-laden construct. Discovering such boundaries, one may glean where humans end, where humans are called into question, and where humans stand to augment themselves or become more than human.

Our understanding about ourselves and about our relationships with nature around us has increased significantly due to the continuous advances in science and technology. Reality is not static since humans and the rest of nature are dynamic, indeed, and both are changing constantly. Transhumanism transcends such static ideas of humanism as humans themselves evolve at an accelerating rate. In the beginning of the twenty-first century, it is now clear than humans are not the end of evolution, but just the beginning of a conscious and technological evolution.

6.3   The Human Seed

Since English naturalist Charles Darwin first published his ideas about evolution on *The Origin of Species* in 1859, it has become clear to the scientific community that species evolve according to interactions among them and with their environment. Species are not static entities but dynamic biological systems in constant evolution. Humans are not the end of evolution in any way, but just the beginning of a better, conscious and technological evolution. The human body might be a good beginning, but we can certainly improve it, upgrade it, and transcend it. Biological evolution through natural selection might be ending, but technological evolution is only accelerating more and more. Technology, which started to show dominance over biological processes many years ago, is finally overtaking biology as the science of life.

As fuzzy logic theorist Bart Kosko has said: "biology is not destiny. It was never more than tendency. It was just nature's first quick and dirty way to compute with meat. Chips are destiny." Photo-qubits might also come after standard silicon-based chips, but even that is only an intermediate means for augmented intelligent life in the universe.

Homo sapiens sapiens is the first species in our planet which is conscious of its own evolution and limitations, and humans will eventually transcend these constraints to become enhanced humans, transhumans and posthumans. It might be a rapid process like caterpillars becoming butterflies, as opposed to the slow evolutionary passage from apes to humans. Future intelligent life forms might not even resemble human beings at all, and carbon-based organisms will mix with a plethora of other organisms. These posthumans will depend not only on carbon-based systems but also on silicon and other "platforms" which might be more convenient for different environments, like traveling in outer space.

Eventually, all these new sentient life forms might be connected to become a global brain, a large interplanetary brain, and even a larger intergalactic brain. The ultimate scientific and philosophical queries will continue to be tackled by these posthuman life forms. Intelligence will keep on evolving and will try to answer the old-age questions of life, the universe and everything. With ethics and wisdom, humans will become posthumans, as science fiction writer David Zindell (1994) suggested:

> "What is a human being, then?"
> "A seed."
> "A… seed?"
> "An acorn that is unafraid to destroy itself in growing into a tree."

Appendix 1 The Principles of Extropy

- **Perpetual Progress:** Extropy means seeking more intelligence, wisdom, and effectiveness, an open-ended lifespan, and the removal of political, cultural, biological, and psychological limits to continuing development. Perpetually overcoming constraints on our progress and possibilities as individuals, as organizations, and as a species. Growing in healthy directions without bound.
- **Self-Transformation:** Extropy means affirming continual ethical, intellectual, and physical self-improvement, through critical and creative thinking, perpetual learning, personal responsibility, proactivity, and experimentation. Using technology—in the widest sense to seek physiological and neurological augmentation along with emotional and psychological refinement.
- **Practical Optimism:** Extropy means fueling action with positive expectations—individuals and organizations being tirelessly proactive. Adopting a rational, action-based optimism or "proaction", in place of both blind faith and stagnant pessimism.
- **Intelligent Technology:** Extropy means designing and managing technologies not as ends in themselves but as effective means for improving life. Applying science and technology creatively and courageously to transcend "natural" but harmful, confining qualities derived from our biological heritage, culture, and environment.
- **Open Society—information and democracy:** Extropy means supporting social orders that foster freedom of communication, freedom of action, experimentation, innovation, questioning, and learning. Opposing authoritarian social control and unnecessary hierarchy and favoring the rule of law and decentralization of power and responsibility. Preferring bargaining over battling, exchange over extortion, and communication over compulsion. Openness to improvement rather than a static utopia. Extropia ("ever-receding stretch goals for society") over utopia ("no place").

- **Self-Direction:** Extropy means valuing independent thinking, individual freedom, personal responsibility, self-direction, self-respect, and a parallel respect for others.
- **Rational Thinking:** Extropy means favoring reason over blind faith and questioning over dogma. It means understanding, experimenting, learning, challenging, and innovating rather than clinging to beliefs.

Appendix 2 The Transhumanist Declaration

1. Humanity stands to be profoundly affected by science and technology in the future. We envision the possibility of broadening human potential by overcoming aging, cognitive shortcomings, involuntary suffering, and our confinement to planet Earth.
2. We believe that humanity's potential is still mostly unrealized. There are possible scenarios that lead to wonderful and exceedingly worthwhile enhanced human conditions.
3. We recognize that humanity faces serious risks, especially from the misuse of new technologies. There are possible realistic scenarios that lead to the loss of most, or even all, of what we hold valuable. Some of these scenarios are drastic, others are subtle. Although all progress is change, not all change is progress.
4. Research effort needs to be invested into understanding these prospects. We need to carefully deliberate how best to reduce risks and expedite beneficial applications. We also need forums where people can constructively discuss what should be done, and a social order where responsible decisions can be implemented.
5. Reduction of existential risks, and development of means for the preservation of life and health, the alleviation of grave suffering, and the improvement of human foresight and wisdom should be pursued as urgent priorities, and heavily funded.
6. Policy making ought to be guided by responsible and inclusive moral vision, taking seriously both opportunities and risks, respecting autonomy and individual rights, and showing solidarity with and concern for the interests and dignity of all people around the globe. We must also consider our moral responsibilities towards generations that will exist in the future.
7. We advocate the well-being of all sentience, including humans, non-human animals, and any future artificial intellects, modified life forms, or other intelligences to which technological and scientific advance may give rise.
8. We favour allowing individuals wide personal choice over how they enable their lives. This includes use of techniques that may be developed to assist memory, concentration, and mental energy; life extension therapies; reproductive choice technologies; cryonics procedures; and many other possible human modification and enhancement technologies.

Bibliography

Asimov I ([1950] 1994) I, Robot. Bantam Books, New York

Bostrom N (2005) A history of transhumanist thought. J Evolut Technol 14(1)

British Telecom (2005) Technology timeline. www.bt.com/technologytimeline. Accessed 4 Aug 2015

Chaisson E (2005) Epic of evolution: seven ages of the cosmos. Columbia University Press, New York

Clarke AC (1984) Profiles of the future: an inquiry into the limits of the possible. Henry Holt and Company, New York

Condorcet N ([1795] 1979) Sketch for a historical picture of the progress of the human mind. Greenwood Press, Westport

Cordeiro JL (1998) Benesuela vs. Venezuela: el combate educativo del siglo. Cedice, Caracas

Cordeiro JL (2010) Telephones and economic development: a worldwide long-term comparison. Lambert Academic Publishing, Saarbrücken

Darwin C ([1859] 2003) The origin of the species. Fine Creative Media, New York

Drexler KE (1987) Engines of creation. Anchor Books, New York. www.e-drexler.com/d/06/00/EOC/EOC_Cover.html. Accessed 4 Aug 2015

Drexler KE (2013) Radical abundance: how a revolution in nanotechnology will change civilization. PublicAffairs, New York

Foundation for the Future (2002) The next thousand years. Foundation for the Future, Bellevue

Fumento M (2003) BioEvolution: How biotechnology is changing the world. Encounter Books, San Francisco

Garreau J (2005) Radical evolution: the promise and peril of enhancing our minds, our bodies—and what It means to be human. Doubleday, New York

Haldane JBS (1924) Daedalus; or, science and the future. K. Paul, Trench, Trubner & Co, London

Haraway DJ (1991) A cyborg manifesto. In: Haraway DJ. Simians, cyborgs and women: the reinvention of nature. Routledge, New York

Hawking S (2002) The theory of everything: the origin and fate of the universe. New Millennium Press, New York

Hughes J (2004) Citizen cyborg: why democratic societies must respond to the redesigned human of the future. Westview Press, Cambridge

Huxley J ([1927] 1957) Transhumanism. In: New bottles for new wine. Chatto & Windus, London

Joy B (2000) Why the future doesn't need us. Wired, April 2000. www.wired.com/wired/archive/8.04/joy.html. Accessed 4 Aug 2015

Kaku M (2012) Physics of the future: how science will shape human destiny and our daily lives by the year 2100. Anchor, New York

Kaku M (2014) The future of the mind: the scientific quest to understand, enhance, and empower the mind. Doubleday, New York

Kurian GT, Molitor GTT (1996) Encyclopedia of the future. Macmillan, New York

Kurzweil R (1999) The age of spiritual machines. Penguin, New York

Kurzweil R (2005) The Singularity is near: when humans transcend biology. Viking, New York

Kurzweil R (2012) How to create a mind: the secret of human thought revealed. Penguin, New York

Minsky M (1987) The society of mind. Simon and Schuster, New York

Minsky M (1994) Will robots inherit the Earth? Scientific American, October 1994. www.ai.mit.edu/people/minsky/papers/sciam.inherit.txt. Accessed 4 Aug 2015

Moravec H (1988) Mind children. Harvard University Press, Boston

Moravec H (1998) Robot: mere machine to transcendent mind. Oxford University Press, Oxford

More M (2003) The principles of extropy. The Extropy Institute. www.extropy.org. Accessed 4 Aug 2015

Mulhall D (2002) Our molecular future. Prometheus Books, Amherst, New York

Paul GS, Cox E (1996) Beyond humanity: cyberevolution and future minds. Charles River Media, Hingham

Roco MC, Bainbridge WS (eds) (2003) Converging technologies for improving human performance. Kluwer, Dordrecht

Roco MC, Montemagno CD (eds) (2004) The coevolution of human potential and converging technologies. New York Academy of Sciences, New York

Sagan C (1977) The dragons of Eden: speculations on the evolution of human intelligence. Random House, New York

Stock G (2002) Redesigning humans: our inevitable genetic future. Houghton Mifflin Company, New York

Teilhard de Chardin P (1964) The future of man. Harper & Row, New York

Venter JC (2008) A life decoded: my genome: my life. Penguin, New York

Venter JC (2013) Life at the speed of light: from the double helix to the dawn of digital life. Viking Adult, New York

Vinge V (1993) The coming technological Singularity. Whole Earth Review, Winter Issue

Warwick K (2003) I, Cyborg. Garnder's, London

Wells HG (1902) The discovery of the future. Unwin, London

World Transhumanist Association (Humanity+) (2002) The transhumanist declaration. http://humanityplus.org/philosophy/transhumanist-declaration/. Accessed 4 Aug 2015

Zindell D (1994) The broken God. Acacia Press, New York

Part II

Foresight for STI Policy: Country Cases

Research Priorities and Foresight Exercises in South Africa: Review and Recent Results

7

Anastassios Pouris and Portia Raphasha

7.1 Introduction

The broad field of futures research has evolved from the United States since the 1950s and 1960s. There is an extensive list of names used in this field interchangeably when referring to futures research such as long range planning, technology assessment, technology forecasting, technology foresight and others. The term "technology foresight" is used in the study.

Irvine and Martin's (1984) seminal work provided one of the first definitions and understandings of foresight and led to a proliferation of relevant exercises.

Foresight took off in the 1990s as European and then other countries began to concentrate their investments in promising areas of science and technology (Martin 1995). Several countries including Japan, the United Kingdom, France, and Germany have undertaken their own large-scale foresight exercises. Some of these countries began to establish relevant organizations with a mandate to inform policy. The practice has spread widely and many developing countries have launched their own foresight exercises.

According to Martin (2002), technology foresight is defined as a process that systematically attempts to look into the longer-term future of science, technology, the economy, the environment and society with the aim of identifying the emerging generic technologies and the underpinning areas of strategic research likely to yield the greatest economic and social benefits.

A. Pouris (✉)
Institute for Technological Innovation, University of Pretoria, Engineering 1 Building, Main Campus, Pretoria, South Africa
e-mail: Anastassios.pouris@up.ac.za

P. Raphasha
Department of Trade and Industry, University of Pretoria, Pretoria, South Africa
e-mail: PRaphasha@thedti.gov.za

© Springer International Publishing Switzerland 2016
L. Gokhberg et al. (eds.), *Deploying Foresight for Policy and Strategy Makers*, Science, Technology and Innovation Studies, DOI 10.1007/978-3-319-25628-3_7

Japan has been one of the leading countries in identifying future technologies since the 1970s. Foresight activities have been institutionalized in the shape of the National Institute of Science and Technology Policy, which is an organization affiliated with MEXT (Ministry of Education, Culture, Sports, Science and Technology) (NISTEP 2010).

Foresight exercises are widely recognized as an appropriate tool in science, technology and innovation (STI) policy design and decision-making processes (Havas et al. 2010). The results of the exercises are often used to identify research priorities, orient policies, and to advise on promising areas for policies. According to Meissner and Cervantes (2008), there is a correlation between the use of technology foresight and a country's innovation performance, indicating that technology foresight has a positive economic impact on a country's innovative potential in the long-term (Meissner and Cervantes 2010). Furthermore, Pietrobelli and Puppato (2015) argue that the successful development trajectories in both Korea and Brazil were partially due to the efforts to link foresight exercises with industrial strategies.

The objective of this article is to review a number of such efforts in South Africa and to report the findings of a recent survey. The recent survey aimed to identify the opinion of relevant stakeholders/industrialists related to the technological needs of the country and to confirm or refute the findings of the dated national foresight exercise.

7.2   Strategic Priority Areas in Technology Development in South Africa: Lessons from the 1990s and 2000s

South Africa has undertaken processes to identify priority areas in technology irregularly. The earliest investigation was undertaken by the Foundation for Research Development (now the National Research Foundation) in the early 1990s (Blankley and Pouris 1993). The investigation first identified the critical technologies of importance that have been developed in other countries. Next, the respondents—representing large companies with their own research and development (R&D) departments, the then South African Scientific Advisory Council, and others—were asked to rate the various technologies by perceived importance.

Figure 7.1 shows the results of the ranking. Over 50 % of respondents identified environmental technologies as being the most important. Computer networks and communication systems followed closely behind (49 %). In third place were software development (42 %) and advanced materials and composites (40 %).

The first official foresight exercise was undertaken by the Department of Arts, Culture, Science and Technology (DACST). DACST undertook and published the National Research and Technology Foresight (NRTF), which was inaugurated in July 1996 and conducted over a 2 year period between 1997 and 1999. The results were published in 2001 (DACST 2001).

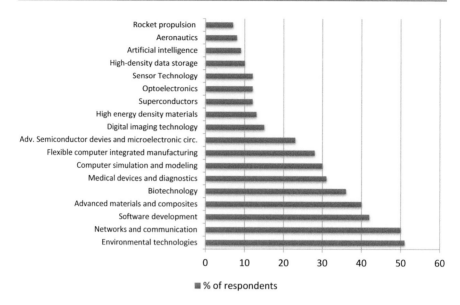

Fig. 7.1 Distribution of critical technologies rated as extremely important by top decision makers: results of a foresight exercise in the early 1990s (share of respondents who rated each technology as most important). *Source*: Blankley and Pouris (1993)

The government intended to utilize the results of the foresight exercise as inputs into government and private sector R&D investment decision-making, and to strengthen the capacity for research in the higher education sector (DACST 1996).

The NRTF study focused on the following sectors of science and technology policy:

- agriculture and agro-processing;
- biodiversity;
- crime prevention, criminal justice and defence;
- energy;
- environment;
- financial services;
- health;
- information and communication technologies;
- manufacturing and materials;
- mining and metallurgy;
- tourism;
- youth.

The concept of the survey was based on a similar work carried out in other countries. Of the 1500 questionnaires distributed to representatives of the

Table 7.1 Evaluation of statements by respondents in the manufacturing sector for the NRTF foresight exercise

No.	Topic	Combined index	WC index	QL index	Constraints
54	Tertiary education institutions (universities and 'technikons') will be transformed to ensure high quality, appropriate skills development that can support a strong manufacturing base	91.94	93.55	90.32	HR, P, F, Soc/Cult
06	Widespread availability of venture capital to enhance the innovation of new products and processes in South Africa	89.29	83.33	82.14	F, P
04	Practical implementation of industry-specific clusters in South Africa to enable the clusters to innovate and compete on world markets	85.63	93.48	77.78	P, HR, F, M
58	Government's appropriate trade and legislative framework will support local industry to meet the challenges of international competition	85.42	87.50	83.33	P
07	Widespread use of intelligent communication systems that will enable SMEs to effectively integrate their skills and knowledge with their chosen industrial partners to form wealth-creating businesses	74.32	78.38	70.27	F, HR, T
60	International transfers and relationship building in the public and private sectors will tangibly help South Africa to use leapfrog technologies to forge ahead	74.32	75.68	72.97	P, HR, F, Soc/Cult
32	Management of new process innovation will be key success factor for most South African companies in the future	74.07	81.48	66.67	HR, T, Infr
08	South Africa's manufacturing production will be predominantly characterized by raw material beneficiation through training of downstream processors on value chain management, design and fabrication technology	72.55	78.43	66.67	HR, T, F
03	Practical application of free-trade zones that will facilitate a regulatory framework for importers and exporters to maintain manufacturing standards in the country of product origin leading to the world economy	69.71	81.08	58.33	P
11	South Africa becomes niche-focused in its manufacturing industry and thus becomes a world leader in a limited number of products	67.27	70.91	63.64	T, HR, Infr, F

(continued)

Table 7.1 (continued)

No.	Topic	Combined index	WC index	QL index	Constraints
09	Mass customization of products, reduced product life cycle, shorter lead times etc. Will become an important driver for South African suppliers to maintain market share on a global basis	65.63	77.08	54.17	T, HR, F
25	Widespread use of practices to eliminate variability in practices and processes is fundamental to competitive manufacturing	61.46	72.92	50.00	HR
61	Widespread use and adherence to international environment and quality standards like ISO9000, ISO14000 and QS9000, VDA6, SABS series, etc. by South African companies to become competitive and internationally recognised	60.29	62.12	58.46	HR, P, F, Soc/Cult
57	In the future, access to mainstream economic and social activities will discriminate between technologically literate and technologically illiterate individuals and groups	58.47	56.41	60.53	HR, Soc/Cult, P, F
12	South Africa's manufacturers develop small-batch manufacturing capabilities for a competitive edge	58.16	65.31	51.02	T, HR, F, M
31	Development of recycling industry (water, raw materials) that will result in waste-free manufacturing	57.14	42.86	71.43	T, P, F
46	Widespread use of concurrent engineering technologies (CIM, CAD, CAM, etc.) to improve time-to-market by South Africa's manufacturing industries	53.41	68.18	38.64	HR, F, T
29	Widespread use of industrial design skills where designer materials will be a fundamental part of new products in the future	53.13	62.50	43.75	HR, T, F

HR human resources, *P* political, *F* financial, *T* technological, *M* market, *Soc/Cult* socio-cultural, *Infr* infrastructure
Source: compiled by authors

manufacturing sector, 150 were returned. The response rate was hence just over 10 %. The relevant committee accepted this was a good response. To analyse the statements, we developed three indices: wealth creation (WC), quality of life (QL), and a combined index (an average of the first two indices). As a representative example, Table 7.1 shows the statements that received the most support from respondents in the manufacturing sector.

Table 7.2 List of future technologies from the 1999 Foresight exercise: short-term horizon

Group of technologies	Components
Continued process and product development of basic materials	• Alloy development; • Polymer development, especially through indigenous coal-based technologies; • Indigenous biomaterials, e.g. natural fibres; • Further processing of precious metals e.g. Platinum group
Downstream product technologies for metal products (e.g. stainless steel, aluminium, precious metals)	• Near-shape processing technologies; • Deeper knowledge and research in optimized technologies for metal forming and joining; • Design and integration of materials in optimum products
Downstream product technologies for polymer products	• Advanced moulding technologies; • Computer-based analysis to support product and process design; • Life-cycle management; • Simulation, modelling and visualization
Computer-based support technologies	• Product design optimization (including virtual prototyping); • Process design and optimization (including plant operation and layout); • Tooling design
Design/product data interchange in value chains	• Development of more energy-conserving processes for raw materials treatment and usage

Source: compiled by authors

The majority of the top-voted statements are of a policy nature. By this we mean that tertiary education institutions, for example, will be transformed to ensure a high quality of appropriate skills, or that industry specific clusters will be created.

The future technologies identified are listed in Tables 7.2 and 7.3 below. The technologies rated as most important were: intelligent communication systems; design and fabrication technology; and concurrent engineering technologies such as CIM, CAD, CAM, etc. The experts considered the least perspective technologies to be: biological structures (biotechnology); semiconductor manufacturing technologies; bio-mimetic systems; 'smart' energy buildings; micro- and nano-technologies for fabrication processes; and ceramic materials for high-temperature gas turbines.

It is worth pointing out that respondents did not consider "futuristic technologies" important. Likewise, the significance of simulation technologies, which are acknowledged worldwide as a cost-effective component of new product and process development, was given limited prominence. The report stated that the typical issues (such as nano-technology and micro fabrication) recommended for future development by foresight exercises of Pacific Rim countries were only given moderate importance; in some cases, they made up some of the ten least important

Table 7.3 List of future technologies from the 1999 Foresight exercise: long-term horizon

Group of technologies	Components
Development of capabilities to implement 'miniaturization' and 'smartness' into products	• Increase precision manufacturing and near-shape technologies; • Direct manufacturing technology ('free-form manufacturing without tooling'); • Integrated sensor/actuator technologies into products
Development of customized materials designed for specific product needs	• Improved methodologies for materials design and development; • Designing with environmentally-friendly/recyclable materials
Development of a manufacturing industry aimed at niche 'information-age' products based on local strengths, despite having essentially missed out on the opportunities of the semiconductor/active materials era in the 1970–1990s	
Introduction off biotechnology development methods to natural fibre optimization for structural composite applications	

Source: compiled by authors

issues. Moreover, the report highlighted the need for South Africa to improve decision making and the development of niche production.

The next large-scale foresight project was carried out in South Africa in 2004 by the Department of Trade and Industry (DTI). The resulting report entitled "Benchmarking of Technology Trends and Technology Developments" study aimed to encompass the following industrial sectors: ICT, tourism, chemicals, biotechnology, automotive industry, aerospace, metals and minerals, culture, clothing and textile, and agro-processing. The study endeavoured to identify global technology development trends, specific current and emerging technologies, and the role of such technologies in sectoral development (DTI 2004).

Within the ICT sector, the most important future technologies were estimated to be wireless network technologies, open source software, telemedicine, and grid computing. In the tourism sector, mobile, environmental, and cultural heritage technologies were considered as highest priority (Table 7.4).

7.3 Strategic Priority Areas in Technology Development in South Africa: Up to 2020

2012 marked a new round of updating the country's industrial technology needs with future planning up to the year 2020, a process carried out by the Department of Trade and Industry. A preliminary list of technologies that are significant for certain sectors up to the year 2020 has been developed based on the experiences of the United Kingdom (Government Office for Science 2010).

Table 7.4 Priority technologies identified in the 2004 Foresight exercise by the DTI

Sector	Technology
ICT	Wireless network technologies
	Home language technologies
	Open source software
	Telemedicine
	Geomatics
	Manufacturing technologies
	Grid computing
	Radio frequency identification (RFID)
Tourism	Mobile technologies
	Wireless technologies
	Internet
	Human languages
	Environmental technologies
	Cultural heritage technologies
Chemicals	Extraction of minerals from coal ash and low value slag
	Fluorine generation and fluorinated organic chemical intermediates,
	New performance chemicals improving the recovery of minerals in the mining sector such as polymer used in solvent extraction processes
	Technologies decreasing economies of scale for chemical plants and hence enabling smaller production facilities to compete against the mega plants
	Low-cost diagnostics and aroma chemicals production
	Development of biodegradable and high-performance polymers
	Bio-diesel and products from alpha-olefins
	Generic pharmaceuticals for meeting future demand for antibiotics and/or anti-retroviral
Biotechnologies	Recombinant therapeutic products and production of generic medicines
	Vaccines against important infectious diseases such as HIV/AIDS, TB, malaria, rotavirus and diarrhoea,
	Diagnostics methods used for screening, diagnosis and monitoring or prognosis of diseases by laboratory methodologies
	Commodity chemicals from biomass
	Energy from renewable resources like plant biomass
	Biocatalysts
Automotive industry	Development of lightweight materials
	Development of alternate fuels e.g. fuel cell technology
	Sensors, electronics and telematics
	Improved design and manufacturing processes
Aerospace	Development of composite materials
	Development of hyper aero-thermodynamics
	Development of sensor usage
	Health and usage monitoring systems
	Noise abatement
	Improved manufacturing processes

(continued)

Table 7.4 (continued)

Sector	Technology
Metal and minerals sector	Light materials extraction
	Alloy technologies, especially in magnesium
	Process improvement
Cultural sector	Product technologies
	Internet
	Online marketing
	Mobile technologies
	Wireless technologies
	Advanced materials
	Human language technologies
	E-commerce
	Environmental technologies
	Portals
Clothing and textile	Intelligent textiles
	High-performance and technical textiles
	Value-added natural fibres—testing systems for foreign fibres in mohair and wool; yarn formation; dying and finishing technologies
	ICT for product and process improvement
Agro processing	Real-time detection of micro organisms in food
	Sensors for online, real-time control and monitoring of food processing
	DNA/RNA chip technologies to speed detection and analysis of toxins in foods
	Food pathogen sensors
	Separation modules that force molecules into confined environments
	Real-time detection systems for verification and validation of intervention technologies used in Hazard Analysis and Critical Control Points (HACCP) systems
	Better understanding of tolerable intake levels for nutraceuticals/dietary supplement components
	Techniques to inactivate micro organisms to yield safer foods with extended shelf lives
	Standardized edible food packaging films
	Biological (e.g. bacteriocins) and chemical inhibitors to prevent or slow growth of pathogens in food
	Technologies for food traceability

Source: (DTI 2004)

7.3.1   Methodology for Selecting Technology Priorities

We developed an open-ended questionnaire related to technology trends and distributed it to several stakeholders, including representatives of key sectors under the remit of the Department of Trade and Industry. The questionnaire was also sent out to researchers possessing close ties with industrial associations, the Technology and Human Resources for Industry Programme (THRIP), and the

Council for Scientific and Industrial Research in South Africa (CSIR), and other organizations.

The response rate was 22 %. Compared to similar exercises, this response rate is satisfactory. The first national foresight exercise in South Africa got a response rate of just 10 %. Most responses came from experts in the chemicals and pharmaceuticals industries (10), followed by the automobile sector (8), textiles, clothing and footwear industry (7), energy (6), and heavy industry (6). The analysis of sectors' current characteristics allows us to draw the following picture. The majority of respondents were in the manufacturing industry, while a smaller proportion was in distribution and assembly. The average age of respondents' companies was 33 years old. Companies employed 900 people on average. Almost two thirds (63 %) of companies declared that they exported their products internationally in sectors such as metallurgical and chemical products, textiles, electronic components and equipment, etc. Approximately a quarter of respondents claimed to be importers, primarily in semi-processed chemical materials, metallurgical rolled stock, power facility and electronic components, medical and pharmaceutical products, etc. Most imported products come from the US and Europe, although a significant share of imports is from Japan, China, and India.

Respondents identified the US, and countries of Europe, Asia, and Africa as potential markets. Respondents said that their companies' turnovers ranged from one million to more than two billion South African Rand (approximately US$77,416–155 million).

7.3.2   Main Results

From the 20 technologies in the list, advanced manufacturing technologies were most often identified as key technologies (58 % of expert respondents). The second most frequently cited key technologies were those connected to modelling and simulation for improving products and processes, reducing the design-to-manufacturing cycle time, and reducing product implementation costs (34 % of respondents). Intelligent sensor network and global computing technologies came in third place (16 %).

The technologies in various sectors that respondents identified as being of most importance at both the current time and in the next 5–10 years are shown in Table 7.5 below.

Table 7.6 below shows the barriers to technological innovation as identified by respondents. The most frequently cited barriers were the high costs for innovation, inadequate funding, and lack of necessary resources. It is noteworthy that more than 50 % of respondents identified a lack of financial resources as a critical barrier.

56 % of responses stated that they acquire technology through their own R&D (Table 7.7). The next most commonly used approaches are by having formal agreements with local companies (13 %) and with foreign companies (12 %). Only 18 % of the companies mentioned that acquire technology through imitation.

Table 7.5 Technologies perceived by respondents as important today and in the next 5–10 years, by sector

Sector	Most important technologies	
	Today[a]	In the next 5–10 years
Aerospace and defence	• Industrial robotics (we are consumers and purchase products from overseas suppliers) • Micro-manufacturing (infancy) • Precision mechanical manufacturing (very important) • Data fusion software (in process) • Infrared optical systems (in process) • Electro-chemical processes • High-speed machining • Additive manufacturing technologies • Space grade sub-systems (in process) • Radar, radio frequency, microwave, electro optics	• Infrared imaging technology manufacturing • Laser communication systems • Embedded software for space systems for radiation tolerant systems • Improved industrial robotics • More energy and eco-friendly systems • Radar, radio frequency, microwave, electro optics
Electronics and ICT	• Biometrics—(limited) • RFID—(limited) • PDA's (available but without local support) • Geographic register for South Africa • Secure and reliable communications • Precision mechanical manufacturing (very important) • Space grade sub-systems (in process) • Linux software development (mid to high importance for free software)	• Biometrics • Infrared imaging technology manufacturing • Laser communication systems • Geographic register for South Africa • Secure and reliable communications • Embedded software for space systems for radiation tolerant systems • Space grade sub-systems
Clothing, textiles, leather and footwear	• Energy efficient processing machinery • Industrial robotics (imported) • Colour physics • Micro-manufacturing (infancy) • Micro-processor controlled machinery with interactive capability	• Flock printing • Coating • Anti-microbe technology • Alternate means of treatment and disposal of factory process effluent • Micro fluidic sensors and diagnostics, lab on a chip • Improved industrial robotics • More energy and eco-friendly systems • Renewable energy
Automobile	• Biotechnology-specific application that are industrially relevant	• Develop further use of polyurethanes • Metal pressing

(continued)

Table 7.5 (continued)

Sector	Most important technologies Today[a]	In the next 5–10 years
	• Stainless steel manipulation • Automation of the manufacturing process • High speed machining • Hybrid Injection moulding machine—advanced • Robot Welding—(available) • Vacuum Forming—(available) • Electro-chemical processes • Powder technology/sintering • Automobile raw material supply chain and value add—(not nearly sufficiently available) • Automobile tier 1 and 2 manufacturing supply upgrade technologies—(not nearly sufficiently available) • International partnerships for technology—(not sufficiently available) • GRP manufacturing processes—(not fully available in South Africa) • Film for covering glass for security and heat load—(not available in South Africa) • Better utilization of available energy resources, including solar energy and fuel cell technology	• Manufacturing expertise for renewable energy • Automobile tier 1 and 2 manufacturing facilities • World class infrastructure manufacturing support • High temperature sintering • Five-axis high speed machining (HSM) • Additive manufacturing technologies • Material technology change • Manufacture of plastic canopies • Polyurethane technology
Agro-processing	• Electronic human interaction platforms (technology available only in imported third and fourth tier end user devices and applications; no visible first or second tier end user support for ICT in the sector) • Modern can and closure manufacturing (status evolving) • Modern metal deck printing technologies • Barrier technologies for safer food storage (not available in South Africa) • Food biotechnology	• Oil stabilisation • Catalysis to upgrade fuel • Water gas shift • Hydrogenation of pyrolysis oils • Modern can and closure manufacturing equipment • Tool and die design and manufacturing • Modern metal deck printing technologies • Emulsifiers • Gasification
Chemicals and pharmaceuticals	• Barrier technologies for safer food storage (not available in South Africa) • Biopolymers, antibacterial	• Biotechnology (industrially relevant applications) • Pyrolysis, oil stabilization, catalysis to upgrade fuel,

(continued)

Table 7.5 (continued)

| Sector | Most important technologies | |
	Today[a]	In the next 5–10 years
	polymers (not available in South Africa) • Sensing and smart polymers— (not available in South Africa) • Advanced process control systems (chemical transformation unit operations) • Powder technology/sintering • Sterile manufacturing • Biotechnology (industrially relevant applications)	gasification, water gas shift • Hydrogenation of pyrolysis oils • Micro fluidic sensors and diagnostics, lab on a chip • Polymers based on bio-sources • Sensing and smart polymers • Automated sterile manufacturing
Creative industries (craft, film, television, music, games etc.)	• IT security • Digital animation • Secure communications • Secure printing (personalized and tamper-proof documents)	• Secure fast internet lines • Visualization of complex data • Secure printing (personalized and tamper-proof documents) • Secure communications • Co-creation tools
Energy	• Renewable solutions, design and manufacture • Small wind turbine design and manufacture • LED lighting technologies • Induction cooking for mainly residential market • Heat pumps water heating high in both residential, commercial and industrial markets • Renewable technologies for mainly residential market	• Small wind technology • LED lighting technologies • Hot water systems • Renewable sources • Improved industrial robotics • Plasma technology, nuclear technology, nanotechnology, mineral beneficiation • Small wind technology • Manufacturing expertise for renewable energy • Better utilization of available energy resources, including solar energy and fuel cell technology
Metallurgy, capital and transport equipment	• Router moulding, plastic injection moulding • Complex brackets using different materials • Robot welding • Casting, forgings manufacturing • On-board computer electronics • Display modules • International partnerships for technology (not sufficiently available) • Automobile tier 1 and 2 manufacturing supply upgrade technologies (not nearly sufficiently available) • Automobile raw material supply chain and value add (not	• Router moulding, plastic injection moulding • Complex brackets using different materials • Robot welding, casting, forgings manufacturing, on-board computer electronics, display modules • World class infrastructure manufacturing support • Automobile tier 1 and 2 manufacturing facilities • Improved industrial robotics • Plasma technology, nuclear technology applications, nano-technology, mineral beneficiation

(continued)

Table 7.5 (continued)

Sector	Most important technologies	
	Today[a]	In the next 5–10 years
	nearly sufficiently available) • Casting • Wear casting—(available) • Electro-chemical processes • High speed machining • Additive manufacturing technologies • Industrial robotics (we are consumers and purchase products from overseas suppliers) • Micro-manufacturing (infancy)	• More energy and eco-friendly systems

[a]The status of technologies stated in brackets is as described by respondents
Source: compiled by authors

Table 7.6 Barriers to technological innovation (share of respondents who chose each option, %)

Barriers to technological innovation	Degree of influence		
	Low	Average	High
Innovation costs too high		10	18
Inadequate funding		11	20
Lack of necessary resources		12	18
Excessive perceived economic risk	4	11	15
Licensing constraints	19	7	2
Lack of qualified personnel	3	15	12
Lack of customer demand for new goods and services	8	14	8
Insufficient flexibility of standards regulation	11	9	10
Organizational inertia within company	8	12	6
Lack of marketing information	12	10	5
Lack of technology information	13	8	6
Lack of cooperation with other firms	12	12	5
Other (specify)			2

Source: calculated by authors

Table 7.7 Acquisition of technologies (share of respondents who chose each option, %)

Undertake own research and development	22
Through formal agreements with companies abroad (e.g. licensing)	12
Through formal agreements with local companies	13
From universities and research councils	10
Through embodied technology in equipment and machinery	9
Through imitation	7

Source: calculated by authors

Table 7.8 Useful policy measures (share of respondents who chose each option, %)

Question: Which policies could help your organization's activity?	
Cluster initiatives	11
Technology platforms	20
Innovation programmes	21
Regulation	10
Competition regulation	5
Quality regulation (labeling, procurement)	8
Fiscal incentives	23

Source: calculated by authors

It should be emphasized that a number of companies mentioned that their research was done abroad.

Table 7.8 shows the policy measures identified by stakeholders as useful for their sectors. The most frequently cited measures were fiscal incentives (23 %), innovation programmes (21 %), and technology platforms (20 %).

Participants of the survey offered several suggestions to promote and support local production, including:

- provide more training on local product development skills;
- boost exports;
- improve skills in fundraising to attract investment;
- make raw materials available at globally competitive prices;
- provide financial and time resources for concept testing;
- liberalize labour laws;
- modernize transport and logistical infrastructure;
- reduce duty exemptions for Southern African Development Community (SADC) countries.

Almost half of respondents (47 %) said they participate in government technology support programmes. Suggested ways to improve such programmes included:

- increase funding for R&D;
- provide funding for purchase of capital equipment;
- increase the salaries for postgraduate students;
- improve the quality of skills and educational programmes;
- respond more quickly to enquiries from business;
- reduce the bureaucracy;
- provide R&D commercialization opportunities for local developers and inventors.

7.4 Conclusion

Over the last three decades, the concept of foresight has become one of the most important tools for priority-setting in science and innovation policy. Typical rationales for foresight exercises have included exploring future opportunities and reorienting science and innovation systems in parallel with building new networks and bringing new actors into the strategic debate (Georghiou and Keenan 2006). It should be emphasized that foresight activities are pursued at different levels, ranging from the organizational to the supranational.

Developing countries or countries with small innovation systems have the potential to benefit from foresight as well as more developed countries. Selectivity is important for these countries, however, as the costs of offering uniform horizontal support to all industrial sectors would be too high and probably not feasible (Lall 2004). Similarly, technologies are not freely available and can only be absorbed if the country is willing to assume the associated costs and risks. Foresight can undoubtedly provide valuable guidance on the above issues.

In South Africa, in contrast to the rest of the world where prioritization exercises are institutionalized and regular, similar efforts are undertaken intermittently and usually result from activities of individual government departments and agencies. This lack of prioritization and coordination of research and innovation agendas has exaggerated the imbalances within the country's system of innovation. Furthermore, industrial enterprises are forced to set up their own technology monitoring mechanisms which leads to substantial diseconomies of scale within the system.

It is interesting to discuss the findings of the most recent survey (2012) in light of the results from the 1999 foresight exercise and international experiences. An important finding of the 1999 foresight exercise was that the participants/stakeholders did not see "futuristic technologies" as important. The most frequently mentioned technologies recommended for future development in foresight processes in the world's leading countries were only given moderate importance in South Africa in 1999 (and in some cases were among the ten least mentioned technologies, e.g. nanotechnology and micro-production). Similarly, the power of simulation technologies, which are acknowledged worldwide as cost-effective components of new product and process development, was given limited prominence. It should be mentioned that the results of the 1999 Foresight appear to have permeated through the scientific and technological system and as a result, the country appears to be lagging in terms of research in emerging technologies (Pouris 2012).

In contrast, the 2012 survey found that stakeholders recognized the importance of emerging and enabling technologies. ICT related technologies (such as secure internet communications, biometrics, robotics, sensors, etc.), biotechnology, and clean energy technologies) were identified by stakeholders as of current importance. Similarly, stakeholders identified "advanced manufacturing technologies", "modelling and simulation for improving products, perfecting processes, reducing design-to-manufacturing cycle time and reducing product implementation costs", and "intelligent sensor network and global computing" as of critical importance for their companies' operations. It should be mentioned that these technologies are at the forefront of priorities internationally. Advanced manufacturing technologies and on-demand manufacturing now attract the attention of most governments around the world in the same way as nanotechnology attracted international support in the early 2000s. The US government is the world leader in terms of allocating substantial resources for advanced manufacturing technologies (Hewitt 2012).

It is clear that the priorities identified during the 1999 Foresight are not necessarily the current STI priorities. In this context, it is important to mention that—in contrast to other countries which monitor and disseminate information related to new technologies—South Africa has no such mechanism. Most countries have institutionalized the monitoring of international priorities and the development of local priorities, and Japan's Foresight are perhaps the most well-known. As discussed here, the lack of South African efforts in the field may be detrimental to the country's manufacturing sector and the performance of its national system of innovation. South Africa has a relatively small national system of innovation with only 0.76 % of gross domestic product spent on R&D (HSRC 2014). Furthermore, the Department of Science and Technology (DST 2015) now seeks to encourage the business sector to spend more on R&D to increase the country's overall R&D expenditures. Foresight among others, may provide the guidance needed by the business sector to fulfil this task.

Acknowledgement   The paper is a revised and developed version of a previously published paper by Pouris A., Raphasha P. (2015) Priorities Setting with Foresight in South Africa. Foresight and STI Governance, vol. 9, no 3, pp. 68–81 with permission from Foresight and STI Governance journal (http://foresight-journal.hse.ru/en/).

References

Blankley OW, Pouris A (1993) Identification of strategic priority areas in technology development. S Afr J Sci 89:169

DACST (1996) White paper on science and technology, preparing for the 21st century. Department of Arts Culture Science and Technology, Pretoria

DACST (2001) Foresight Synthesis Report: Dawn of a New Century. Department of Arts Culture Science and Technology, Pretoria

Department of Trade & Industry (2004) Benchmarking of technology trends and technology developments. Pretoria

DST (2015) Strategic plan for fiscal years 2015–2020. Department of Science and Technology, Pretoria

Georghiou L, Keenan M (2006) Evaluation of national foresight activities: assessing rationale, process and impact. Technol Forecast Soc Chang 73:761–777

Government Office for Science (2010) Technology and innovation futures: UK growth opportunities for the 2020s. Foresight Horizon Scanning Centre, Government Office for Science, http://www.bis.gov.uk/assets/bispartners/Foresight/docs/general-publications/10-1252-technology-and-innovation-futures.pdf

Havas A, Schartinger D, Weber M (2010) The impact of foresight on innovation policy-making: recent experiences and future perspectives. Res Eval 19(2):91–104

Hewitt K (2012) The future of U.S. manufacturing—a literature analysis (Part III). MIT Washington Office, Washington, DC

HSRC (2014) R&D survey statistical report 2012/13. Human Sciences Research Council, Pretoria

Irvine J, Martin BR (1984) Foresight in science. Pinter, London

Lall S (2004) Reinventing industrial strategy: the role of government policy in building industrial competitiveness, G-24 Discussion paper, 28, United Nations

Martin BR (1995) Foresight in science and technology. Technol Anal Strat Manag 7(2):139–168

Martin BR (2002) Technology foresight in a rapidly globalizing economy. In: International practice in technology foresight. United Nations Industrial Development Organization, Vienna

Meissner D, Cervantes M (2008) Results and impact of national foresight-studies. Paper presented at the third international Seville seminar on future-oriented technology analysis: impacts and implications for policy and decision-making, Seville, Spain, 16–17 Oct

Meissner D, Cervantes M (2010) Successful Foresight study: implications for design, preparatory activities and tools to use. Foresight-Russia 4(1):74–81

NISTEP (2010) The 9th science and technology foresight—contribution of science and technology to future society—the 9th Delphi survey. Science and Technology Foresight Centre National Institute of Science and Technology Policy, Tokyo

Pietrobelli C, Puppato F (2015) Technology foresight and industrial strategy in developing countries. UNU-MERIT working paper series 2015–2016, Maastricht

Pouris A (2012) Science in South Africa: the dawn of a renaissance? S Afr J Sci 108(7/8):66–71

STI Foresight in Brazil

8

Cristiano Cagnin

8.1 Introduction

In essence, the objective of foresight is to shape spaces for structured dialogue that fosters engagement, creativity and reflection, both individual and collective. Hence, the aim is to use the future as a trigger to spark imagination and expand our understanding of the present through structured conversation to collectively imagine the future and make choices in the present (Miller 2007, 2011a, b).

A number of methods, tools, instruments and techniques are used to structure dialogue and shape possible future developments. However, critical in the design and implementation of a foresight process is the comprehension of the relation between context, content and approach (Cagnin et al. 2008). Moreover, expected results and associated impacts, both tangible and intangible, should be defined from the outset (Da Costa et al. 2008).

Foresight approaches have evolved through successive generations or phases, which are not mutually exclusive (Johnston 2002, 2007; Cuhls 2003; Georghiou 2001, 2007): (1) technology forecasting or internal dynamics of technology, with participation of experts; (2) interaction between technology and markets, with participation happening across the academic-industry nexus; (3) interaction between markets and social actors, with an user-oriented perspective and broader societal participation; (4) distributed role in the science and innovation system, with multiple organisations carrying out exercises fit for individual purposes but coordinated with other activities; and (5) mix of distributed exercises focused on either structures or actors within the science, technology and innovation (STI) system, or on the scientific/technological dimensions of broader social and economic issues and challenges.

C. Cagnin (✉)
Center for Strategic Studies and Management in Science, Technology and Innovation, SCS Qd 9, Lote C, Torre C, 4° andar, Salas 401 A 405, Ed. Parque Cidade Corporate, Brasília-DF CEP 70308-200, Brazil
e-mail: ccagnin@cgee.org.br

© Springer International Publishing Switzerland 2016
L. Gokhberg et al. (eds.), *Deploying Foresight for Policy and Strategy Makers*,
Science, Technology and Innovation Studies, DOI 10.1007/978-3-319-25628-3_8
113

Foresight practice occurs mainly in two 'modes', although a combination of both is possible and becoming commonplace. In 'mode 1' the aim is to improve or optimise the existing system (Weber 2006; Eriksson and Weber 2006; Havas et al. 2007). 'Mode 2', on the other hand, focuses on debating and promoting fundamental changes of established paradigms (Da Costa et al. 2008). At the same time, a number of principles guide foresight work (adapted from Keenan et al. 2006): (1) future orientation in the medium to long term; (2) active participation of stakeholders; (3) use of evidence and informed opinions, combining thus interpretation and creative approaches; (4) coordination; (5) multidisciplinarity; and (6) action-orientation.

Globally, advanced countries and institutions practice a combination of generations four and five as well as 'modes' 1 and 2. This takes place routinely and with close attention to the six principles mentioned above. The aim is to increase the relevance of foresight activities and its impacts in the decision making process, such as in the design and implementation of public policy. CGEE is, therefore, aiming to advance in this direction rather than concentrating efforts only in generations one to three and in 'mode 1' foresight.

8.2   Foresight Evolution

In the post-industrial revolution, which caused a great deal of social and technological transformations, a sense of preoccupation towards the future became more widespread. During this time the attention was on the improvement of decision processes and public debate, and the focus was on the anticipation of trends and long-term implications of short-term decisions.

In the nineteenth and twentieth centuries classical economists centred their analysis on the future of capitalism economies. In the beginning of the 1900s the principles of trends extrapolation and social indicators were established. The term foresight appeared in a speech delivered by H. G. Wells for the Royal Institution in 1902 entitled "The discovery of the future", were the thesis that the future could be known or understood scientifically was defended. The first systematic methods of experts' analysis were developed towards the second half of the twentieth century (e.g. Delphi and cross impact analysis) as well as the first simulation studies.

In the 30s and 40s, after WWI and under the effects of the Great Depression, a new world order looked at S&T as a means to redemption. H. G. Wells published "An experiment in Prophecy" anticipating the world in 2000 by predicting modern transport dispersing people from cities into suburbs, moral restrictions diminishing due to sexual freedom, and the formation of the EU. Wells also defended in 1932 the institutionalisation of what he called "departments and professors of foresight". In 1945 a committee had the task to look ahead 20 years to tell where the aviation sector was evolving to and how the US Air Force would get there. Future studies initiated towards the second half of the 40s when institutions like RAND and SRI were created to develop long-term planning analysing systematic trends for military purposes soon after the WW2.

In the 50 and 60s, after WW2 and the establishment of the Cold War, the focus of future studies reduced to the anticipation of future technologies, mainly for defence objectives. RAND and SRI used system analysis and developed the games theory, scenarios and Delphi. The focus was on S&T and engineering developed by and for military application and big corporations. A limited number of experts and futurists were involved in these activities, and the main methods used were Delphi, scenarios, brainstorming and expert panels. The conceptual and methodological basis of foresight was developed in this period. Hence, this is considered to be the birth of foresight modern practice based on operational research efficiency and aiming at deliberate interventions to direct desired change. The main concern was on probabilistic analysis of what may happen in the future based on an extrapolation of what happened in the past (i.e. forecasting). Key work in this period is: "The art of conjecture" (de Jouvenel 1972) and "Inventing the future" (Dennis Gabor 1963). In 1966, the first future-oriented university course was developed in the US by Alvin Toffler at The New School (New York). This is considered to be the practice of mode 1 and first generation foresight.

During the 70s the world understands the limits of forecasting due to oil crises and the failure of predictions such as "Limits to Growth" (Meadows et al. 1972) and the Bariloche Foundation, "Catastrophe or New Society?" (1976). Unpredictable events lead into a wider understanding that global systems are uncertain and complex. Forecasting becomes less deterministic and 'accepts' that the future is not a mere extension of the past and that discontinuities do occur. Japan uses forecasting methods about the future of S&T to inform its policies, including in its analysis social and economic needs as well as advances in S&T. A number of activities start worldwide such as the Futuribles Project in France, the Committee for the Next 30 Years in the UK and the Hudson Institute in the US (a spin-off of RAND). Projects oriented to socio-political objectives and methods that provide guidance and fundamentals to analyse alternative situations and choices, such as scenarios, gain importance. GE and Shell start using scenarios to support its strategic decisions, and in 1976 Shell looks ahead to 2000 identifying discontinuities in the industry. After the oil crises (1974) almost half of the firms listed on Fortune 1000 use foresight techniques in its planning processes. The same occurs in Europe. This is considered to be the practice of mode 1 and second generation foresight.

In Brazil, the 70s is considered to be the "embryonic phase" of foresight (Porto 2012). Theoretical and methodological work is initiated in the country towards the end of the 70s. Henrique Rattner releases the book "Future Studies—Introduction to technological and social anticipation". The first formal group to think long-term (prospectively) on S&T policy is established in 1979 at Unicamp by Amilcar Herrera. The first official explicit document on S&T policy is released as part of the Development National Plan (I PND, 1972–1974): the Basic Plan of S&T Development (I PBDCT). However, only the II PBDCT integrated into the II PND (1974–1979) brought about innovations such as the creation of the National System of S&T Development (SNDCT) and of the National Programme of Post-

Graduation (PNPG). The latter demonstrated for the first time a harmony between a national plan and that of S&T (Salles-Filho and Corder 2003).

In the 80s exercises worldwide consider in its analysis multiple futures that embrace global and social uncertainties. In 1983 the term foresight is connected to S&T at SPRU and in 1985 Michael Godet developed the school La Prospective. Institutional foresight gains attention of national governments as an activity associated with the identification of long-term priorities and the development of S&T policies. Activities developed in France (National Colloquium on Research & Technology) and in Holland (Ministry of Education and Science) are good examples (van Dijk 1991). This is considered to be the practice of mode 1 and third generation foresight.

In Brazil, the 80s is considered to be the "emergency phase" of foresight (Porto 2012). Scenarios start been used in the second half of the decade by governmental companies that operate in long-term sectors (Buarque 1998), such as energy. Examples of this are the BNDES (development bank) embedding scenarios in its strategic planning process around 1984, Eletrobrás/Eletronorte (energy firm) in 1987, and Petrobrás (oil company) in 1989 to analyse market and demand for energy and fuel. In fact, Petrobrás initiated the use of scenarios together with BNDES in 1986. In 1987 CENPES (research branch of Petrobrás) developed its first technological scenarios, and in 1989 scenarios became an intrinsic part of its strategic planning. Scenarios also had influence in both the business and the academic environments. The results of the "scenarios for the Brazilian economy—competitive integration" proposed an update of the country's industrial structure, an open and competitive economy, and the renegotiation of Brazil's external debt in the long run and in better conditions. This was realised in the government of president Collor in the 90s. Also, the creation of the National Council of S&T (CCT) in 1985 influenced the rebirth of future thinking in Brazil, although its fragile institutional setting (initially subordinated to SEPLAN/PR) and the excessive preoccupation with a short-term agenda lead to the discontinuation of a long-term planning. The ministerial management of S&T in the period known as the New Republic improved financial and operational aspects but did not fix the coordination deficiencies. This is considered to be the practice of mode 1 and a mix of first and second generation foresight (while the world is practicing already the third generation).

In the 90s foresight exercises become widely organised by governments, advisory groups, research advisors, national academies of sciences and other governmental departments worldwide, as well as by industrial associations and firms. Large scale programmes take place in Germany, France and the UK, which inspired other EU and OECD countries, as well as Latin America and Asian countries (notably Japan, Korea, China and India) to initiate their own national programmes. S&T were the central foci of these activities that aimed at identifying strategic areas of research and emerging technologies that could reap economic (competitiveness) and social (visions, networks, education and culture) benefits. This is considered to be the practice of mode 1 and forth generation foresight.

In Brazil, the 90s is considered to be the "dissemination phase" of foresight (Porto 2012; Massari 2013). EMBRAPA (governmental food research firm) adopts a long-term approach in its strategic planning. The agribusiness and value chains become important concepts for a more systemic understanding embedded in future analysis. The creation of a new CCT (National Council of S&T) establishes two boards: (1) prospective, information and international cooperation, and (2) regional development. The first board enabled an in-depth debate around the future of the national S&T system leading to yet another rebirth of future thinking and its embeddedness into the public sector. Themes like future technologies and the role of information as a transformative instrument gained attention. In 1997 a study was proposed emulating the French Key Technologies project, aiming at identifying technological priority topics of S&T in sectorial themes. The objective was to orient the decisions of CCT as well as to involve the Ministry of S&T and the public sector in thinking about the future in order to define future priorities and strategies. In 1998 the project Brasil 2020 initiated at SAE was the first governmental experience in undertaking an integrated planning for the country in recent years. It aimed at fostering a reflection about which country Brazil would like to be and what was needed to transform this vision in a reality. Workshops and interviews generated inputs for scenarios and a broad consultation of social actors tried to grasp societal aspirations. Equity, justice and quality of life were central aspects of society's hopes and ambitions; all still valid today. This is considered to be the practice of mode 1 and a mix of second and third generation foresight (while the world is practicing the fourth generation).

As society's complexity increases worldwide, from the year 2000 onwards the scope and focus of foresight activities enlarges to cover a diversity of themes. Foresight exercises shift the emphasis from scope and coverage to pay more attention in the process. Methods start been used with more criteria and according to context. Foresight activities adapt into a world with greater complexity, inter-connectivity and interdependencies. These activities try to answer Grand Challenges and the needs for sustainability of public policy in an adaptable way. The understanding of complex systems and possible future behaviours of social actors become the departing point and the focus become challenges instead of decision-making silos. Coordination of societal actors to solve common problems is sought for, and foresight is institutionalised in Australasia (Australia, Korea, China, Taiwan, Singapore, etc.) beyond the EU and Japan amongst other countries. This is considered to be the practice of a mix of modes 1 and 2 as well as of fourth and fifth generation foresight.

In Brazil, from the year 2000 onwards is considered to be the "continuous dissemination and generalisation phase" of foresight (Porto 2012; Massari 2013). The sectoral funds and a movement initiated by the Ministry of Science, Technology and Innovation (STI) lead to a revolution in STI in the beginning of the decade. However, these have been partially discontinued in the last years. Nonetheless, the seeds that germinated from the CCT resulted in the creation of the ProspeCTar programme (Ministry of STI) and, to a certain degree, in the Brazilian Programme of Prospective Industrial Technology (PBPTI) within the Ministry of Development,

Industry and Commerce (MIDIC) in partnership with UNIDO. Delphi was the main technique used. The project "tendencies" of MSTI and MDIC supported by the Sectorial Fund of Oil and Gas aimed at a wide understanding of trends ahead for the sector with a 10 years timeframe. The methodology embraced scenarios, diagnosis, desk research, text mining, expert panels, webdelphi, among other methods. The project "strategic directives" (DECTI) resulted, in 2001, in the Second National STI Conference and in the creation of CGEE in order to institutionalise foresight and evaluation studies nationally. According to Santos and Fellows-Filho (2009), other results from the Second National STI Conference were the publication of the Green (showing the STI trajectory over the last 50 years together with transformative initiatives and future opportunities) and White (showing the STI lines that should comprise the national STI policy over the next 10 years—towards 2012—in order to consolidate a national STI system) books on STI.

The project "Brasil 3 Times" (NAE/PR) aimed at defining strategic long-term objectives for the country and to build a pact between the State and society in order to achieve these objectives, beyond trying to institutionalise a long-term vision in the public strategic management. Scenarios were the main method employed. Embraer (aviation firm) uses scenarios and Delphi routinely and, more recently, simulation systems in order to detect emerging signals. Technology foresight in Brazil is used as an instrument to formulate STI public policies with a focus on sectors and value chains. However, in spite of all mentioned activities, the results do not attain the expected impacts as it did in other countries worldwide. Aulicino (2006) observes that possible failures reside in the ways in which these exercises were formulated, designed and executed. According to him, all had a little degree of public participation. Also, according to the author, there was a lack of understanding of concepts, objectives and expected impacts in these exercises, which lead to little engagement and sharing of ideas between social actors, as well as the absence of new networks that were expected as a result. This is considered to be the practice of mode 1 and a mix of first to third generation foresight.

In this context, foresight in Brazil is still marked by the dichotomy between discontinuity and the institutionalisation of activities that can become embedded explicitly in decision-making and planning processes. At the same time, the focus needs to shift from technology alone to innovation more broadly in order to identify and articulate anticipatory intelligence that serves to reorient the NIS systemically, thus embracing social, environmental, economic, political, technological and behavioural (values) aspects. Coordination between decision-making silos (i.e. ministries) and social actors (fostering broad societal participation) still needs to be promoted more widely and with a focus on challenges or common problems. Moreover, fostering dialogue and participation instead of stakeholders' consultation alone is important to attain a more systemic understanding of the challenges at hand as well as to build commitment of individual actors to collective decisions. Finally, promoting these changes means that there is a need to shift the focus of foresight activities from optimisation alone to one that builds a bridge between optimisation and contingency at the same time that it embraces uncertainty, complexity and creativity.

8.3 Orienting the National Innovation System Through Foresight[1]

In recent years, the ways in which NIS can be reoriented to address Grand Challenges have been widely debated. According to Cagnin et al. (2012), these are challenges which are complex and difficult or even impossible to solve by single agencies or through rational planning approaches alone. Academics and activists have understood this for some time and the articulation of these challenges is not new. The novelty here relies in the increasing attention given to such issues in the formulation of national STI policies. The reasons for this are complex. In part, it reflects the increasing perception of urgency in responding to a series of challenges that can have devastating consequences at local/global scales in the next decades if neglected. But it also reflects an attempt to redirect STI efforts, at least those financed by the public sector, to respond explicitly to political agendas. The central question is how to support such a mission focused on challenges for an innovation practice (Freeman 1970; Rogers 1995; Freeman and Soete 1997; OECD and Eurostat 2005; Fagerberg et al. 2004; Hall and Rosenberg 2010) to develop which is more directed and transformative through the use of foresight methods and approaches (Cagnin et al. 2012).

Foresight processes and approaches offer decision makers the potential to look through disruptive transformations which are necessary as a solution to or caused by Grand Challenges. From the perspective of transcending epistemological and ontological barriers to better respond to Grand Challenges, foresight connects long-term perspectives and different knowledge bases into the decision making process. In doing so it puts emphasis in multiple and holistic approaches under which it is possible to identify diverse triggers and instruments to shape the direction of innovation systems. These processes also help in the use and management of the uncertainties associated with the activities and functions of innovation systems (Bach and Matt 2005; Bergek et al. 2008; Edquist 2008; Hekkert et al. 2007; Jacobsson and Bergek 2006; van Lente 1993; von Hippel 2005; Woolthius et al. 2005), as well as with the future more widely. It does so through the creation of spaces for social, economic and political actors to meet and appreciate their positions vis-a-vis possible future directions of innovation (Cagnin et al. 2012).

From the political perspective, this potential of coordination improves the communication and the understanding between different decision-making silos giving support, therefore, to the emergence of an effective combination of policies that fosters innovation. Finally, the simple fact of participating in such processes can in itself be transformative by encouraging the adoption of new perspectives and the development of new abilities to detect and process weak signals of change. In this way, different approaches and processes can enable actors to become more adaptive and capable of realising systemic changes. In order to do so, foresight can act different roles to orient innovation systems so that these are better able to

---

[1] Cf. Cagnin et al. 2012.

respond to Grand Challenges (Cagnin et al. 2012). These roles can be grouped in (Barré and Keenan 2008; Da Costa et al. 2008; Cagnin et al. 2011, 2012): inform the decision making process, structure and mobilise networks of actors, and enable innovation system actors.

8.4   Foresight at CGEE

CGEE's mission is to promote STI to advance economic growth, competitiveness and well being in Brazil. It does so by carrying out foresight and strategic evaluation studies in combination with information and knowledge management approaches and systems. At the core of its activities is its position and ability to articulate and coordinate diverse actors within the Brazilian National Innovation System (NIS). One of the CGEE's institutional objectives linked to its mission is to lead foresight studies that generate anticipatory intelligence to the Brazilian NIS in general and to the Ministry of STI and its agencies in particular.

The institution is changing its approach to developing and addressing new strategic questions, and in recognising new issues which merit further investigation via systemic and systematic observations and dialogue. It is doing so in order to evolve its foresight practice to combine generations one to five as well as 'modes' 1 and 2 (see introduction), and to enable its results to be better positioned to support reorienting the Brazilian NIS.

In this context, CGEE is undertaking a transformative process by changing its approach to design, organise, implement, manage and evaluate its foresight studies. The aim is to move from a normative and prescriptive approach to one that embraces complexity, emergence and novelty. Such a move is being sought by fostering an improvement in the institution's capability to use systematic approaches and to develop recommendations for policy design and implementation based on shared insights and perceptions. The institution is moving in this direction in order to improve the quality and robustness of its anticipatory intelligence and to increase the preparedness of the NIS for disruptive events (Cagnin 2014). CGEE is attaining this objective via the creation of spaces for dialogue between key players from different domains, with diverging views and experiences. These spaces are designed to develop vision- and consensus-building processes for considering and inducing "guided" processes of transformation, as well as to shape and define dialogues on likely transformations and policy discussions on tackling major changes, and on research and innovation agendas. A number of tools and approaches are been explored to enable the institution to advance in this direction and to use the future to ignite and expand the collective imagination and understanding of the present. It is important to highlight that the approach been developed by CGEE is considering three integrating themes[2] that determine the quality of foresight processes:

---

[2] The role of methods is to provide support to these three integrating themes.

- Expertise (i.e. to understand the nature of the problem/challenge at hand, to recognise emergence and substantive patterns from weak signals of change in a noisy environment and from collective distributed intelligence).
- Creativity (i.e. in the art of embracing "know knowns", "known unknowns", "unknown knowns" and "unknown unknowns", thus considering knowledge, opinions, speculations and conjectures. Also, in the ability to imagine, to experiment and to interpret novel and transformative possibilities of the future in the present, in the ability of embracing the emergent future, and in the ability to tell stories through narratives and visualisation).
- Interaction or alignment (i.e. government, science and industry, policy makers and politicians, which requires both mental and physical handshaking).

Therefore, the aim of foresight at CGEE is to balance contextualised design with systemic and systematic qualitative and quantitative approaches, and to welcome unknowability and uncertainty as a source of novelty, thus also providing an invitation to creativity and improvisation. Working with possible, probable, desirable, plausible and reframed[3] futures provides a way to work with unknowable futures and novel frames for imagining the future. According to Miller (2007, 2011a, b)), it does so by exposing anticipatory assumptions and revealing the social processes and systems used to invent and describe imaginary futures. The author affirms that such processes increase our capability to imagine discontinuity and to put more effort into inventing what is unknowable, thus developing greater capacity to use the future; what he calls futures literacy.

Developing the above mentioned balance implies building an ability of "walking on two legs"[4]: improve or optimise the current system at the same time as it moves towards new and/or disruptive system configurations. Being able to operate both in known systems (inside-in, inside-out, and outside-in) with more efficiency and efficacy as well as to operate in unknown systems (outside-out), according to Fig. 8.1, will support the institution in crafting strategic questions for itself and its clients. In other words, looking outside the system in which we are familiar with will support not only developing and addressing new strategic questions, but also in recognising new issues (e.g. challenges, technologies, social transformations, etc.) through systematic observations and dialogue, and selecting those which are worth further investigating in order to identify new opportunities.

In short, optimisation focuses on the improvement of existing systems and looks at the future detached from the present. It usually allows for incremental innovation based upon a normative future with prescriptive actions associated. It prepares one to operate in known systems or 'inside-in' which, in other words, means that the

---

[3] Futures Literacy (Miller 2011a, b).

[4] Presentation delivered by Riel Miller in the Futures Literacy UNESCO Knowledge Labs (FL Uknowlab) or Local Scoping Exercises (LSE).

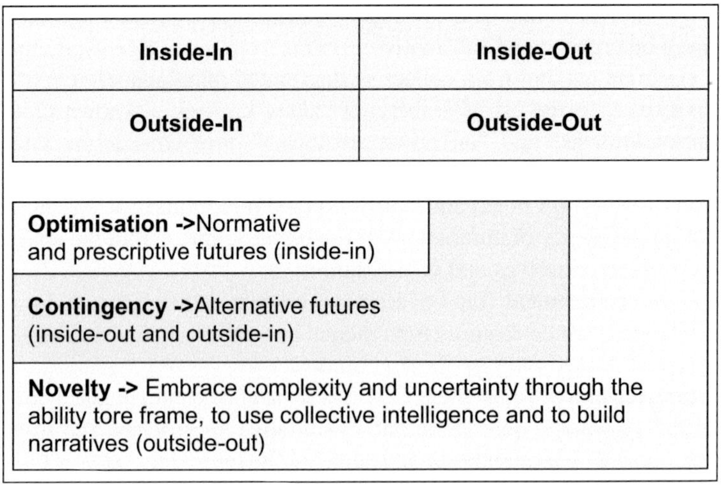

Inside-In	Inside-Out
Outside-In	Outside-Out

Optimisation ->Normative
and prescriptive futures (inside-in)

Contingency ->Alternative futures
(inside-out and outside-in)

Novelty -> Embrace complexity and uncertainty through the
ability tore frame, to use collective intelligence and to build
narratives (outside-out)

Fig. 8.1 Operating both in known and unknown systems. *Source*: adapted from Miller (2007–2011a, b)

boundaries of the system are well understood and only what resides within such boundaries is analysed.

Contingency, on the other hand, focuses on avoiding something undesirable to happen or on preparing the current system to continue to exist in the future. It also looks at the future detached from the present, but looks at alternative futures instead of looking at one single vision alone. The aim is to enable one to prepare for different possibilities of the future regardless of these becoming or not a reality, as well as to shape a desirable pathway with checkpoints that once monitored enables one to adapt to new events or situations along the way. Here beyond looking 'inside-in' (within known systems) it enables one to look both 'inside-out' and 'outside-in' the system under analysis. In other words, it enables one to identify how changes in the system being analysed (therefore known, at least partially) can impact other systems and vice versa. Innovation promoted here is also incremental but with potential to foster more radical or disruptive innovation.

Being able to embrace complexity and uncertainty, however, means putting a stronger focus on narratives and in the ability to reframe (questions, concepts, cultures, etc.) our images and metaphors about the future. According to Miller (2011a, b) this means that the future is not detached from the present but is an alternative intrinsic part of it, which enables us to embrace the 'unknown' and the unexpected in the present while the future unfolds. The focus in on more than one transformative future ('outside-out') that is open to discontinuity as well as to birth and rebirth. At the end such an approach allows for both incremental and radical or disruptive innovation, with experimentation being at the heart of our capacity to cultivate and reap the new and the unexpected (Miller 2011a, b).

Based on the above, the direction in which foresight is evolving at CGEE aims to enable the institution to operate at all above systems in parallel. In doing so it invites uncertainty, complexity and creativity throughout the process and translates these into actual recommendations for policy design and implementation or into new strategic questions that should be investigated and addressed in order to reorient the Brazilian NIS.

References

Aulicino AL (2006) Foresight para políticas de CT&I com desenvolvimento sustentável: estudo de caso Brasil, 306 f. Tese (Doutorado em Administração)—Departamento de Administração da Faculdade de Economia, Administração e Contabilidade,Universidade de São Paulo, São Paulo

Bach L, Matt M (2005) From economic foundations to S&T policy tools: a comparative analysis of the dominant paradigms. In: Matt M, Llerena P (eds) Innovation policy in a knowledge based economy: theories and practices. Springer, Heidelberg, New York, Dordrecht, London, pp 17–40

Barré R, Keenan M (2008) Revisiting foresight rationales: what lessons from the social sciences and humanities? In: Cagnin C, Keenan M, Johnston R, Scapolo F, Barré R (eds) Future-oriented technology analysis. Springer, Heidelberg, New York, Dordrecht, London, pp 41–52

Bergek A, Jacobsson S, Carlsson B, Lindmark S, Rickne A (2008) Analyzing the functional dynamics of technological innovation systems: a scheme of analysis. Res Policy 37:407–429

Buarque SC (1998) Experiências recentes de elaboração de cenários do Brasil e da Amazônia brasileira. Parcerias Estratégicas 5:1–26

Cagnin C (2014) STI foresight in Brazil. Foresight-Russia 8(2):46–55

Cagnin C, Keenan M, Johnston R, Scapolo F, Barré R (eds) (2008) Future-oriented technology analysis: strategic intelligence for an innovative economy. Springer, Heidelberg, New York, Dordrecht, London

Cagnin C, Loveridge D, Saritas O (2011) FTA and equity: new approaches to governance. Futures 43:279–291

Cagnin C, Amanatidou E, Keenan M (2012) Orienting European innovation systems towards grand challenges and the roles that FTA can play. Sci Public Policy 39:140–152

Cuhls K (2003) From forecasting to foresight processes—new participative foresight activities in Germany. J Forecast 23:93–111

Da Costa O, Warnke P, Cagnin C, Scapolo F (2008) The impact of foresight on policy-making: insights from the FORLEARN mutual learning process. Technol Anal Strat Manag 20 (3):369–387

De Jouvenel B (1972) The art of conjecture. Transaction Publishers, Piscataway, NJ

Edquist C (2008) Design of innovation policy through diagnostic analysis: identification of systemic problems (or failures). CIRCLE Electronic Working Paper Series 2008/06. Lund University, Lund

Eriksson EA, Weber M (2006) Adaptive foresight: navigating the complex landscape of policy strategies. Second international Seville seminar on future-oriented technology analysis, Seville, 28–29 Sept

Fagerberg J, Mowery DC, Nelson RR (2004) The Oxford handbook of innovation. Oxford University Press, Oxford

Freeman E (1970) Stakeholder theory of the modern corporation. In: Hoffman M, Frederick RE, Schwartz MS (eds) Business ethics: readings and cases in corporate morality, 4th edn. McGraw-Hill, New York

Gabor D (1963) Inventing the future, vol 663. Secker & Warburg, London

Freeman C, Soete L (1997) The economics of industrial innovation, 3rd edn. Pinter, London

Georghiou L (2001) Third generation foresight—integrating the socio-economic dimension. In: International conference on technology foresight—the approach to and the potential for new technology foresight. Science and Technology Foresight Center, National Institute of Science and Technology Policy (NISTEP), Ministry of Education, Culture, Sports, Science and Technology, Japan

Georghiou L (2007) Future of forecasting for economic development. In: UNIDO Technology Foresight Summit, Budapest, 27–29 Sept

Hall BH, Rosenberg N (2010) Handbook of the economics of innovation. Elsevier, Amsterdam

Havas A, Schartinger D, Weber KM (2007) Experiences and practices of technology foresight in the European region. In: UNIDO Technology Foresight Summit, Budapest, 27–29 Sept

Hekkert MP, Suurs RAA, Negro SO, Kuhlmann S, Smits REHM (2007) Functions of innovation systems: a new approach for analysing technological change. Technol Forecast Soc Chang 74:413–432

Jacobsson S, Bergek A (2006) A framework for guiding policy-makers intervening in emerging innovation systems in 'catching-up' countries. Eur J Dev Res 18:687–707

Johnston R (2002) The state and contribution of international foresight: new challenges. In: The role of foresight in the selection of research policy priorities, JRC-IPTS Seminar, Seville, 13–14 May

Johnston R (2007) Future critical and key industrial technologies as driving forces for economic development and competitiveness. In: UNIDO Technology Foresight Summit, Budapest, 27–29 Sept

Keenan M, Butter M, Sainz de la Fuenta G, Popper R (2006) Mapping foresight in Europe and other regions of the world. The 2006 Annual Mapping Report of the EFMN, European Foresight Monitoring Network (www.efmn.eu)

Meadows DH, Meadows DL, Randers J, Behrens WW (1972) The limits to growth. Universe Books, New York, NY, p 102

Miller R (2007) Futures literacy: a hybrid strategic scenario method. Futures 39:341–362

Miller R (2011a) Futures literacy—embracing complexity and using the future. Ethos 10:23–28

Miller R (2011b) Being without existing: the futures community at a turning point? A comment on Jay Ogilvy's "Facing the fold". Foresight 13(4):24–34

OECD and Eurostat (2005) Oslo manual: guidelines for collecting and interpreting innovation data, 3rd edn. OECD, Paris

Porto C (2012) Prospective foresight in Brazil: an overview and cases. Mutual Leaning Workshop on Scenarios, Brazil, CGEE, 5 Dec

Rattner H (1979) Estudos do futuro: introdução à antecipaçãotecnológica e social. FGV, Rio de Janeiro

Salles-Filho S, Corder S (2003) Reestruturação da política de ciência e tecnologia e mecanismos de financiamento à inovação tecnológica no Brasil. Cadernos de Estudos Avançados 1:35–43

Rogers EM (1995) Diffusion of innovations, 4th edn. Free Press, New York

Santos DM, Fellows-Filho L (2009) Prospectivana América Latina: Evolução e desafios. Canal6, Bauru

Van Dijk TA (1991) The interdisciplinary study of news as discourse. A handbook of qualitative methodologies for mass communication research. pp 108–120

van Lente H (1993) Promising technology, the dynamics of expectations in technological development. PhD thesis, University of Twente

von Hippel E (2005) Democratising innovation. MIT Press, Cambridge

Weber M (2006) Foresight and adaptive planning as complementary elements in anticipatory policy-making: a conceptual and methodological approach. In: Voss JP, Bauknecht D, Kemp R (eds) Reflexive governance for sustainable development. Edward Elgar, Cheltenham, pp 189–221

Woolthius RK, Lankhuizen M, Gilsing V (2005) A system failure framework for innovation policy design. Technovation 25:609–619

National System of Science and Technology Foresight in Russia

9

Alexander Chulok

9.1 Determinants of S&T Foresight Systems

Economies around the world are facing significant challenges arising from globalised value chains, emerging markets and production facilities, changing societies' attitudes and demands as well as accelerating science and technology (S&T) progress among others. These challenges however allow new opportunities taking shape in all sectors of economy including high-tech manufacturing and services as well as traditional sectors. Also these challenges offer promising potential through satisfying newly emerging economic and societal demands, rapid replacement of outdated technologies and deployment of new technological solutions on a massive scale. In this respect searching for breakthrough areas is a prerequisite to meeting country's economic competitiveness and welfare ambitions (Bhagwati 2004; Meissner 2012). Foresight has been widely used mainly for engaging key stakeholders (business, academia, government, society et al) in building shared visions of the future of countries and respective economies and societies and to identify key S&T areas capable to generate corresponding products and services and achieve seamless integration into existing global value chains, and the creation of new ones (Georghiou et al. 2008).

Among the obstacles to keep pace with these developments at country level is the still persistent gap between science and the real economy (including the

A. Chulok (✉)
Institute for Statistical Studies and Economics of Knowledge, National Research University Higher School of Economics, 20 Myasnitskaya Street, 101000 Moscow, Russia
e-mail: achulok@hse.ru

© Springer International Publishing Switzerland 2016
L. Gokhberg et al. (eds.), *Deploying Foresight for Policy and Strategy Makers*,
Science, Technology and Innovation Studies, DOI 10.1007/978-3-319-25628-3_9

manufacturing sector), which is articulated in a variety of communication patterns of the different communities involved in the respective field. In many instances, business lacks a comprehensive view of the new paths of technological development which at the same time is common knowledge in the scientific community (RAND 2006; Meissner et al. 2013). The scientific community in turn needs a better understanding of which technologies may be demanded in the future. Exchange between science and business (with active participation by the government) supports the creation of a shared vision of the future of socio-economic, scientific and technological development and communication improves understanding of future needs of the economy and society (Sokolov and Chulok 2012). It allows businesses to be better aware of the complex world of modern technology and adjust their long-term strategies accordingly (Meissner 2014).

Many countries realise that the implemented portfolio of science, technology and innovation (STI) policy instruments needs to be reconsidered through the prism of "Grand Responses to Grand Challenges" paradigm—this shift could be seen through the changing agendas of national strategic documents (European Commission 2010a, b; KISTEP 2012; NISTEP 2010a, b). New global challenges are also caused by fundamental changes in the manufacturing, the transformation of socio-economic processes, and shifts in cultural values, and as a result, the redistribution of profit centres in global value chains (Perez 2002). New frequently unforeseen markets and niches are emerging, while traditional, previously profitable production techniques are quickly being sidelined (McKinsey 2013; OECD 2012). Innovations permeate every sector of the economy, and many assumptions underpinning economic models that have traditionally been the foundation for management decisions are becoming irrelevant today (Vishnevskiy et al. 2015).

The evolution of information technologies, whose pervasive role is yet to be realised, and globalisation processes have drastically expanded opportunities for social interaction and interpersonal communication. This has changed developed countries' system of values associated with the development of a knowledge economy substantially. With the scale of networked communications growing rapidly, economic and societal requirements and requests for S&T development are increasingly interdisciplinary in nature, whereas the lines between different disciplines are blurring (Fig. 9.1). The interdisciplinarity and hyperconnectivity of these communications and developments require the improved coordination of STI policy instruments to implement a consistent and coherent policy mix.

Developing an evidence based STI policy mix suitable to respond to these challenges poses new tasks for S&T foresight and anticipatory STI policy including an integration of quantitative and qualitative methods and approaches for collecting evidence and ultimately becoming part of the policy making and implementation process. Therefore open and transparent communication and respective networking are a precondition to assure the appropriate engagement and support of a broad range of stakeholders. Moreover it needs to account for the fact that the main basic assumptions of established and also modern economic models are changing dramatically, such as the rational behavior of households, shifting economies of scale

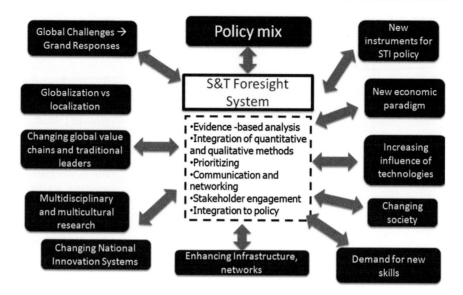

Fig. 9.1 New challenges for STI policy. *Source*: National Research University Higher School of Economics

together with the classical structure of value chains and business models among others accordingly.

Thus the basis for an anticipatory STI policy mix needs to be changed compared to the existing approaches. The experience of many developed and rapidly growing countries demonstrates that a better wired-up national S&T foresights designed to identify promising areas of S&T that ensure technological leadership, contributes to economic growth and social development, and setting the appropriate policy priorities is essential. Therefore numerous countries have developed approaches, which connect foresight with STI policy (Table 9.1).

However there is no 'one fits all model' for the incorporation of S&T foresight in national STI policy, on the contrary the objectives and motivation and also the mechanisms of including foresight in policy making vary strongly between countries (Gokhberg 2016). This is not only apparent in the title of final documents which build on the respective foresight activities but also in the underlying scope and consequently the respective STI policy measures developed and implemented. Moreover the range and number of knowledge holders, e.g. experts involved in the S&T foresight activities vary significantly, e.g. in China more than 1200 experts are involved,[1] in South Korea between 2000 and 5000,[2] in Japan between 1300 and 4300[3] and in India between

---

[1] http://www.nistep.go.jp/achiev/ftx/eng/mat077e/html/mat077de.html.

[2] 1st, 2nd round of national foresight.

[3] Delphi surveys.

Table 9.1 Key players of national S&T foresight: selected countries[a]

Country	Key governmental stakeholders	Key foresight centres (examples)	Number of S&T foresight activities (estimation)	Integration into STI policy (examples)
China	• Ministry of Science and Technology of the People's Republic of China (MOST)	• Chinese Academy of Science and Technology for Development (CASTED)	>20	• Selection of National Critical Technologies (1995)[b, c] • Technology Foresight of Priority Industries in China (1999)[d] • The National Medium- and Long-Term Programme for Science and Technology Development (2006–2020)[e] • State High-Tech Development Plan (863 Programme)[f, g, h] • Integrated National Technology Roadmap
Republic of Korea	• National Science and Technology Commission (NSTC) • Ministry of Science, ICT and Future Planning (MSIP) • National Research Foundation (NRF)	• Korea Institute of S&T Evaluation and Planning (KISTEP) • Korea Institute of R&D Human Resources Development (KIRD)	>15	• Vision 2025: Korea's Long-term Plan for Science and Technology Development[i] • National Technology Roadmap[j] • National 5-year plans of S&T development[k] • Strategic initiatives in the most important technology fields • Basic plans for S&T development
Japan	• Ministry of Education, Culture, Sports, Science and Technology (MEXT)	• National Institute of Science and Technology Policy (NISTEP) • Science and Technology Foresight Center	>25	• Science and Technology Basic Plans[l, m] • Japan's Innovation Strategy 2025[n] • Strategic priorities for the key S&T areas

(continued)

Table 9.1 (continued)

Country	Key governmental stakeholders	Key foresight centres (examples)	Number of S&T foresight activities (estimation)	Integration into STI policy (examples)
	• Council for Science and Technology • Management and Coordination Agency • Japan Science and Technology Agency, Center for Research and Development Strategy			• National 5-year plans of S&T development • Innovation 25: Creating the Future, Challenging Unlimited Possibilities[o]
United States	• US President's Council of Advisors on Science and Technology (PCAST) • US Department of Defense (DoD) • US Department of Energy (DoE) • Office of Science and Technology • Institute for the Future • RAND corporation	• Defense Advanced Research Projects Agency (DARPA) • Massachusetts Institute of Technology (MIT) • US Mack Center for Technological Innovation • National Renewable Energy Laboratory • US National Science Foundation (NSF)	>100	• Critical technologies • Innovation strategies • National technology initiatives • Sectoral strategic plans • Innovation and S&T development priorities
Turkey	• Ministry of Science, Industry and Technology • Turkish State Planning Organization	• Scientific and Technological Research Council of Turkey (TUBITAK)	<10	• Vision 2023: Turkish National Foresight[p]

(continued)

Table 9.1 (continued)

Country	Key governmental stakeholders	Key foresight centres (examples)	Number of S&T foresight activities (estimation)	Integration into STI policy (examples)
Brazil	• Ministry of Science, Technology and Innovation • Ministry of Development, Industry and Foreign Trade	• Center for Strategic Studies and Management in Science, Technology and Innovation (CGEE) • FINEP Research and Innovation	>10	• Scenarios of the Brazilian Economy, BNDES (1984–90)[q] • Scenarios about the future of the Amazon Region (ELETRONORTE) (1988, 1998)[p] • Introduction and Consolidation of the Use of Scenarios in Petrobras (1989, 1992)[p] • Scenarios in the State of Minas Gerais (2003–2007)[p]
India	• Government of India, Department of Science and Technology	• Technology Information, Forecasting and Assessment Council	<10	• Technology Vision for India for 2020[r] • Technology Vision for India for 2035[s, t]
European Union[u, v, w, x, y]	• European Commission (EC) • European Foresight Monitoring Network (EFMN) • European Foresight Platform (EFP) • DG Joint Research Centre • Foresight and Behaviourial Insights Unit	• European Commission's Joint Research Centre • Institute for Prospective Technological Studies	>550	• Strategy for European Technology Platforms: ETP 2020[z] • Strategic Research Agendas of the European Technology Platforms • Horizon 2020 Research and Innovation Programme[aa, bb, cc, dd, ee]

(continued)

Table 9.1 (continued)

Country	Key governmental stakeholders	Key foresight centres (examples)	Number of S&T foresight activities (estimation)	Integration into STI policy (examples)
The Netherlands[ff]	• Ministry of Education, Culture and Science • Foresight Committee of the Royal Netherlands Academy of Arts and Sciences • Centraal Plan Bureau (CPB) • Sociaal Cultureel Planbureau (SCP)	• Netherlands Organization for Applied Scientific Research (TNO)	>150	• 2025 Vision for Science: Choices for the Future[gg]
United Kingdom[hh, ii, jj]	• Department for Business, Innovation and Skills (BIS)	• University of Manchester Institute of Innovation Research	>100	• Technology and Innovation Futures: UK Growth Opportunities for the 2020-s[kk]
Germany[ll]	• Federal Ministry of Education and Research	• Fraunhofer Institute for Systems and Innovation Research	>30	• High-Tech Strategy 2020 for Germany[mm]

Source: National Research University Higher School of Economics
[a]The author would like to express gratitude to his colleague Ilya Kuzminov for noticeable help in constructing this table
[b]http://www.nistep.go.jp/achiev/ftx/eng/mat077e/html/mat077de.html
[c]http://ictt.by/eng/portals/0/FS2009_TF20China160309.pdf
[d]http://www.researchgate.net/publication/242477676_Technology_foresight_and_critical_technology_selection_in_China
[e]http://sydney.edu.au/global-health/international-networks/National_Outline_for_Medium_and_Long_Term_ST_Development1.doc
[f]http://www.pacificcommunityventures.org/uploads/misc/case_studies/13-High_Tech_RD_Program.pdf
[g]http://erawatch.jrc.ec.europa.eu/erawatch/opencms/information/country_pages/cn/supportmeasure/support_mig_0009
[h]http://www.most.gov.cn/eng/programmes1/
[i]http://unpan1.un.org/intradoc/groups/public/documents/APCITY/UNPAN008040.pdf
[j]http://www.nistep.go.jp/IC/ic030227/pdf/p5-1.pdf
[k]http://unpan1.un.org/intradoc/groups/public/documents/APCITY/UNPAN008049.pdf
[l]http://www.jsps.go.jp/english/e-quart/38/JSPS38-L.pdf
[m]http://erawatch.jrc.ec.europa.eu/erawatch/opencms/information/country_pages/jp/policydocument/policydoc_0011

[n]http://japan.kantei.go.jp/innovation/interimbody_e.html
[o]http://www.cao.go.jp/innovation/en/pdf/innovation25_interim_full.pdf
[p]Saritas O, Taymaz E, Tumer T (2006) Vision 2023: Turkey's national technology foresight program—a contextualist description and analysis. ERC Working Papers in Economics 06/01
[q]http://macroplan.com.br/Documentos/NoticiaMacroplan2012127192922.pdf
[r]http://tifac.org.in/index.php?option=com_content&view=article&id=52&Itemid=213
[s]http://tifac.org.in/index.php?option=com_content&view=article&id=835&Itemid=1403
[t]http://www.tifac.org.in/index.php?option=com_content&view=article&id=863&Itemid=1402
[u]EC (2011) iKNOW ERA toolkit. Applications of wild cards and weak signals to the grand challenges and thematic priorities of the European research area
[v]EC (2011) Monitoring foresight activities in Europe and the rest of the world. Final report
[w]UNIDO (2005a) Technology foresight manual. Volume 1. Organization and Methods. United Nations Industrial Development Organization, Vienna
[x]UNIDO (2005b) Technology foresight manual. Volume 2. Technology Foresight in Action. United Nations Industrial Development Organization, Vienna
[y]COM White Paper European Governance, EU 2001
[z]ftp://ftp.cordis.europa.eu/pub/etp/docs/swd-2013-strategy-etp-2020_en.pdf
[aa]Popper R, Velasco G, Edler J, Amanatidou E (2015) ERA open advice for the evolving dimensions of the European research and innovation lands. Policy brief. Manchester Institute of Innovation Research, Manchester
[bb]http://ec.europa.eu/programmes/horizon2020/sites/horizon2020/files/H2020_inBrief_EN_Final BAT.pdf
[cc]http://ec.europa.eu/regional_policy/sources/docgener/guides/synergy/synergies_en.pdf
[dd]http://ec.europa.eu/research/horizon2020/pdf/press/horizon2020-presentation.pdf
[ee]https://ec.europa.eu/research/participants/data/ref/h2020/wp/2014_2015/main/h2020-wp1415-swfs_en.pdf
[ff]Mapping foresight revealing how Europe and other world regions navigate into the future (2009) EFMN, Luxembourg
[gg]http://www.government.nl/documents-and-publications/reports/2014/12/08/2025-vision-for-science-choices-for-the-future.html
[hh]Havas A, Schartinger D, Weber M (2010) The impact of foresight on innovation policy-making: recent experiences and future perspectives. Research Evaluation 19 (2):91–104
[ii]Mapping foresight revealing how Europe and other world regions navigate into the future (2009) EFMN, Luxembourg
[jj]Miles I (2010) The development of technology foresight: a review. Technological Forecasting and Social Change 77:1448–1456
[kk]https://www.gov.uk/government/uploads/system/uploads/attachment_data/file/288564/10-1252-technology-and-innovation-futures.pdf
[ll]Mapping foresight revealing how Europe and other world regions navigate into the future (2009) EFMN, Luxembourg
[mm]http://www.bmbf.de/pub/hts_2020_en.pdf

500 and 5000.[4] These numbers reflect the range of subject fields and topics covered as well as the shape and connectedness of the NIS itself.

S&T foresight and future-oriented technology analysis could provide "'informing', 'structuring' and 'capacity-building' benefits while enabling a shift

---

[4] http://www.tifac.org.in/index.php?option=com_content&view=article&id=914&Itemid=1412.

in innovation foci towards Grand Challenges" (Cagnin et al. 2012). Governments increasingly have to moderate and orchestrate national S&T foresight instead of actively intervening. In its turn S&T foresight needs to develop organically and build a potential to react flexibly to changing requirements and framework conditions. Accumulation of "critical mass" of foresight projects, teams, skills, and demand from stakeholders can serve as a bridgehead to create a fully-fledged S&T foresight system—an institutional framework which sets up conditions for coordinated systemic foresight activities.

Furthermore S&T foresight systems urgently have to reflect the individual characteristics and features of the NIS. In this regard it is important to consider the nature of the institutional communication between key stakeholders in the national innovation system when designing the approach to building a relevant national S&T foresight system. The communication paths and styles depend substantially on nation-specific features, including the organization and structure of the national economy and S&T in particular.

Specific S&T foresight studies are aimed at specific issues meeting key challenges for particular NISs. Accordingly for countries with less experience in S&T foresight it shows challenging to directly copy an existing system for S&T foresight due to differences in established economic agents and communication routines in the respective NIS. Hence each country has to develop and implement its own special S&T foresight system oriented at the established interaction routines of agents in the country. However this does not prevent countries from learning from other international experiences.

S&T foresight systems institutionalize foresight activities in countries which aim at leveraging national STI competences, assuring continuous methodological development and integration in STI policy mix design and implementation.

9.2 The Russian S&T Foresight System

Since the 1990s, Russia has seen a powerful surge in interest in areas of long-term S&T foresight at different levels, including national (Gokhberg and Kuznetsova 2011a, b; HSE 2013) and sectoral (Chulok 2009). Moreover, while foresight studies initially came "from the top", e.g. from the government (federal ministries and agencies, development centres, etc.), recent activities have multiplied at the regional level, mostly in industrially developed Russian regions and cities, such as Moscow, Yekaterinburg, Samara, Bashkortostan, Krasnoyarsky, Sakha (Yakutiya), etc. Also a significant share of S&T foresight work is conducted by major businesses in course of preparing their R&D strategies and programmes. Russia has already accumulated considerable experience of implementing more than 50 long-term strategic projects, namely national and industry-specific S&T foresight studies, strategies for specific sectors of the economy, strategic research programmes of technological platforms, programmes of regional innovation clusters, innovation programmes of state-owned companies, etc. (Sokolov 2009; Makarova and Sokolova 2014).

In this regard the Russian Ministry of Education and Science initiated creating and updating a system of S&T priority areas and a list of the national critical technologies on the basis of a long-term S&T foresight. Significant methodological groundwork has also been performed by other actors: Ministry of Industry and Trade, Ministry of Communications and Mass Media, Ministry of Natural Resources, Ministry of Health, State Corporation Rosatom, Federal Space Agency ("Roskosmos"), Skolkovo Foundation, et al. These activities have also substantially contributed to the involvement of foresight in the development of strategic STI policy programmes to establish technology platforms and programmes to develop innovation clusters. The latter are in particular expected to influence the of industry-specific and regional priorities for the medium- and long-term with respect to research and development, and aligning tools for collaboration between science and industry.

The largest state-owned companies are creating and implementing innovation development programmes, which are medium-term action plans designed to develop and deploy new technologies and world-class innovative products and services, and advance the innovative development of key branches of industry. In preparing those programmes the companies typically also take advantage of broader S&T foresight activities that therefore can be considered as an instrument providing a fertile ground for the innovation development of the economy.

Russia has accumulated a "critical mass" of strategic projects, skills, and knowledge, which made it possible to formally establish an S&T foresight system in Russia (Fig. 9.2) in fulfillment of the Decree of the President of the Russian Federation No. 596 of 7 May 2012 "On the Government's Long-Term Economic Policy". This system's creation was also complemented by The Federal Law "On Government Strategic Planning in the Russian Federation" adopted in June 2014.

The development of the S&T foresight system assumes that strategic information will be generated for decision making at national, industrial, regional, and corporate levels of governance.

a) The system's results at the national level are presented in the medium- and long-term forecasts of socio-economic development, strategic foresight of the risks of socio-economic development and threats to national security, framework for the long-term development of the Russian Federation's long-term socio-economic development, and basic areas of activity for the Government in the medium-term.

b) At the industry level, the S&T foresight system encompasses government programmes, industry-specific strategic planning documents, strategic research programmes of technology platforms, innovation strategies and programmes for major state-owned companies among others.

c) Regional medium- and long-term socio-economic foresights and programmes to develop innovation clusters created at the regional level.

The outputs produced by the S&T foresight system should be used to develop, implement, and update the indicated government strategic planning documents,

Fig. 9.2 National S&T foresight activities: key official milestones. *Source*: National Research University Higher School of Economics

some of which are already in effect, while others should be drawn up in the nearest future (Message from the President of the Russian Federation to the Federal Assembly 2012). As an integral part of government strategic planning, this system should be focused on supplying key stakeholders' current and future needs for strategically important information on the most significant trends in global innovation development, the prospects of appearance of new products and key manufacturing technologies necessary for their creation. Therefore long-term S&T foresight; priority S&T areas and critical technologies; industry specific and technology roadmaps become the functional backbone of the S&T foresight system at the national level.

9.3   Long-Term S&T Foresight

The Russian long-term S&T foresight outlines a high-quality picture of expected technological breakthroughs and the most important S&T areas, which bring promising scientific results, key technological developments, and innovative products (services). It was developed as a series of consecutive cycles, connected both in terms of the priority area development horizon being estimated and the methodology being used covering a time horizon of 20 years or even more. Additionally, all long-term exercises under the S&T foresight system are closely aligned with the creation and refinement of priority areas and critical technologies.

Taking into account different product life cycles (from 4–5 years in information and telecommunication sector to 15–20 in aircraft or shipbuilding), the key element of is a long-term foresight. The foresigth exercise should the transition to the task-oriented and functional priorities that will respond to global challenges and the most urgent national problems. It emphasizes interdisciplinary areas and projects and includes the establishment of a permanent network of experts, including Russian and foreign experts.

9.4 Priority S&T Areas and the List of Critical Technologies

Priority S&T areas define landmarks for the S&T development promising the greatest contribution to national security, accelerated economic growth, and increased national competitiveness through developing the economy's technology base and high-tech industries. Priority areas are specified in the list of critical technologies and the most important promising innovative products, which to a large degree will determine the growth of traditional markets and emergence of new ones.

Selecting and revising priority S&T areas and critical technologies is closely interrelated with the long-term S&T foresight since they are implemented in a sequence (one after another) and provide an extensive information background for each other.

The results of S&T foresight as well as S&T priority areas and the list of critical technologies are among the key sources of information for designing government S&T-related programmes, identifying the structure and content of the most important government-funded innovation projects, planning the development of research infrastructure, and STI policy regulations.

9.5 Roadmaps for Individual Industries and Inter-Industrial Clusters

The objective of roadmaps is to establish connections between future demand for innovative products and technological solutions and the ability to actually create them. Using a roadmap facilitates the construction of a long-term sequence of steps to satisfy existing and emerging demand in various segments. A roadmap offers a clear visual representation as to which life cycle stage individual products and technologies will belong in the future. Roadmaps guide strategies for market entry and technology development as well as R&D programmes required to create the technological solutions demanded by the market. On top of that, roadmaps have to propose alternative ways to achieve the established goals. Roadmaps present a comprehensive, interconnected view of future developments of key technologies and products in individual sectors of the economy and society. They enable the shift from identifying promising areas (markets, products, technologies, research areas) to proposing specific innovation projects and describing the key steps required to implement them.

Long-term S&T foresight represented in the roadmap are focused on development of policy and information support for the corresponding decision-making processes. Roadmaps also provide recommendations regarding the organisation of government support for the most important STI programmes and projects. The S&T foresight system assumes the creation of national roadmaps among others for primary S&T areas and key sectors of the manufacturing industry. The roadmaps for S&T areas describe paths to achieve crucial scientific results that will enable future breakthrough technologies and products. They all reflect strategic R&D programmes that will lead to accomplishing the roadmap's stated goals. Sectoral roadmaps describe possible paths to the development of a specific group of technologies and new products. They also indicate potential alternatives and depict future trends of technological development, areas of application for technologies, and market trends influencing innovative products' functional characteristics.

The national S&T foresight is complemented with industry-specific foresight, industry-specific S&T priorities and critical technologies, as well as regional foresight studies and roadmaps. These activities use methodologies similar to those at the national level although they primarily consider markets, products, and technologies linked to an industry or region, and are focused on potential use in industry-specific and regional programmes, as well as governmental agency programmes for particular industries.

9.6   Institutional Structure of the S&T Foresight System

Building an effective S&T foresight system requires, above all, the design and development of an organisational structure with clearly defined properties (objectives, functions, responsibilities) for its constituent parts. The entities responsible for activities pertaining to the system's creation and operation should also be clearly defined. Russia's S&T foresight system was established with the participation of key ministries and governmental agencies responsible for the national development, the Academy of Sciences, and economic development agencies (Fig. 9.3).

The S&T foresight system is supported by a dedicated expert unit based at the National Research University Higher School of Economics, which also employs and Advisory Board with representatives from leading international foresight centres from the UK, the USA, Korea, China, Canada among others. The main role in the creation, development, and coordination of S&T foresight system's activities designed to supply the manufacturing sector's future needs, including the development of key technologies are assigned to the Interministerial Commission on Technology Foresight of the Presidium of the Russian Federation's Presidential Council on Modernisation of the Economy and the Innovative Development of Russia (Minutes No. 1 of 28 June 2013) (Foresight 2014). The Interministerial Commission has the following functions:

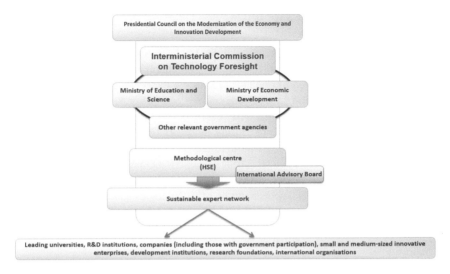

Fig. 9.3 Institutional organisation of the national S&T Foresight system in Russia. *Source*: National Research University Higher School of Economics

1. Preparation of proposals regarding *coordination of and support for federal authorities'* and other interested parties' activities to create the S&T foresight system with respect to using the results of S&T foresight studies (both ongoing and earlier completed) in Russia as well as coordination of objectives, tasks, responsible entities, expected results, and implementation time frames of S&T foresight studies. Furthermore the Commission supports the creation of unified methodologies to ensure the operation of the foresight system, including the allocation of federal authorities' and other interested parties' functions with respect to creating and supporting the operation of the S&T foresight system.

2. Support the *operation of the S&T foresight system* with respect to monitoring the operation of the system, analysis of its outputs and preparation of analytical and information materials based on them regarding the state and future of Russia's S&T development given the changing socio-economic situation. In addition it assures the use of the foresight system's outputs in government strategic planning of the Russian Federation's socio-economic development and in revising the S&T priority areas and the list of national critical technologies.

3. Preparation of proposals to promote increased *communication and coordination* activities of federal authorities and other stakeholders to create and support the operation of the foresight system; reporting of the system's outputs to stakeholders, expansion of their use to modernise the economy and advance Russia's innovation development.

To ensure the implementation of these functions, the Interministerial Commission includes representatives of federal authorities, national academies of sciences, research organisations, universities, and businesses. The organisational and technical support of the Commission is provided by the Russian Ministry of Education and Science.

These results of S&T foresight studies are used in preparing and implementing STI policy tools:

- laws and regulations concerning S&T development;
- frameworks and strategies for medium- and long-term development; long-term socio-economic forecast government programmes and federal special-purpose programmes; industry-specific strategies;
- programmes for innovation development of large state-owned companies;
- other strategic planning documents.

S&T foresight provides also a wide range of information to increase efficiency of development and implementation of strategic initiatives in the field of S&T and innovation.

The key S&T foresight system's operational tasks are:

- development of a long-term national S&T foresight;
- compilation and updating of national priority S&T areas and the list of critical technologies;
- development of industry-specific S&T foresight studies;
- S&T priorities and lists of critical technologies;
- organisation and annual monitoring of S&T development;
- development of roadmaps for sectors of the economy;
- preparation of proposals to use the foresight system's results in government strategic planning documents;
- coordination of and support for foresight activities performed by federal authorities and other stakeholders;
- creation and maintaining of the national pool of experts;
- development of communications platforms to discuss and use the foresight results;
- improvement of the methodology and development of common standards for studies performed in the framework of the S&T foresight system;
- creation of a unified publicly-available database of foresight-related materials.

However, there is still has a number of problems and constraints hindering the effective creation and implementation of long-term foresight studies, due to certain interagency discoordination and governance gaps.

1. *Insufficient coordination of sectoral foresight activities.*

The majority of Russia's foresight activities on the development of manufacturing technologies is financed by sectoral ministries and agencies and, to a smaller degree, economic development agencies and major companies.

They are mostly aimed at solving particular sector-specific problems and in very rarely cover broader interagency issues. This leads to considerable duplication of activities by different agencies. Insufficient attention is paid to major inter-industry problems.

2. *Lack of unified methodological approaches to S&T foresight exercises.*

Many foresight projects have implemented novel methodological approaches, some of which account for the best international practices, but in general the methodological problems of organising foresight activities are still to be further developed. A coordinated methodological approach to a wide spectrum of problems is needed: determination of foresight studies' objectives, time horizons, organisational solutions, choice and scope of problem domains and their level of detail, which methods should be applied and in what sequence, involvement of experts, content and formats for presenting the resulting documents, among others. Also the absence of generally accepted approaches or long-standing practices regarding the identification of industry-specific S&T priorities is critical. Whereas at the national level there are established methods of identifying S&T priority areas and critical technologies, at the industry level they still have to be developed. Unified methodological solutions must provide a framework for use by not only federal ministries and agencies, but also by economic development agencies, major companies, technology platforms, regional innovation clusters and other entities involved in foresight activities. Another serious problem is the lack of minimal quality standards for S&T foresight. In recent years, many studies have pretended to be based on foresight methodology. However, they were not such neither by their methodological approach nor by the depth of research. It significantly discredits modern approaches to foresight and, in the eyes of potential users, reduces the value of foresight results produced by teams involving a large number of highly-qualified experts.

3. *Organisational problems in building an S&T foresight system, including managing the use of foresight results by different actors.*

As noted above, a specific legal and regulatory framework has been formed for organizing the S&T foresight system. However, even with this framework in place, actual implementation of S&T foresight is of considerable importance. The first step is to arrange a series of measures designed to organise interaction between the various actors of the system, prepare "minimal quality standards" for foresight studies based on the best national and international practices, and determine communication processes and formats (negotiation at an early stage of research programmes, involvement of experts, information exchange, etc.).

4. *Creating a unified network of S&T foresight centres within the S&T foresight system.*

The last few years have brought substantial progress in the creation of a network of S&T foresight experts. In 2011, under the auspices of the Russian Ministry of Education and Science and coordinated by the Higher School of Economics, a network of industry-specific foresight centres, built around leading

Russian universities,[5] was created for S&T priority areas. This network presently includes more than 200 research organisations, universities and companies in 30 regions of Russia and more than 15 foreign countries. However, the number of communication platforms where S&T foresight outputs can be discussed is still insufficient. Mechanisms for reporting them to all stakeholders have not yet been fully established.

Attempts to solve these problems with individual foresight studies developed by different ministries and agencies are not always insufficient. However, it must be stated that the "critical mass" of foresight projects that Russia has accumulated up to now form a solid foundation for successfully creating a national S&T foresight system.

9.7 Conclusion

For the last decade, national STI policies have increasingly been challenged by the quest for anticipatory policy approaches, which overcome the long established reactive nature of policy instruments. While for a long time the STI policy had been designed to assure global competitiveness of countries the focus recently shifted to building awareness of future developments of economies and societies and designing relevant STI policy instruments. From this viewpoint the policy "toolkit" should be able to provide rapid response to opening windows of opportunities and coming threats that emerge from changing landscape of global trends and challenges. Development of strategic STI policy tools is a big challenge in particular for developing countries. Among the countries' response to this challenge is the establishment of S&T foresight systems providing support to evidence-based anticipatory STI policy.

Hence S&T foresight in its different forms and varieties has become an integral part of these STI policy making systems. Recently a demand for the coordination of policy-making and policy implementation by various ministries and agencies (so-called "Whole-of-government innovation policy") has been widely used by OECD countries. This rising demand requires the development of a "smart methodology" of foresight exercises which should combine quantitative and qualitative methods and timely adjustment to the needs of decision-makers. Another aspect of this process could be noticed in changing culture and "style" of governance in STI policy: most countries that have long foresight history are moving from fully-fledged large-scale national exercises to more targeted and specified problem-oriented activities; while "newcomers" start from exploring the full landscape first.

---

[5] Pursuant to a decision of the Interministerial Commission on Technology Foresight to regularly update the Russian long-term S&T Foresight (Minutes from 29 October 2010), as part of creating a network of industry-specific S&T Foresight centres built around the country's leading universities.

The case of Russia shows an attempt of creating a national S&T foresight system as an important and logical step toward building a more elaborated future oriented STI policy. The transformation of the Russian economy is intended to transition from a catch-up model to raising Russias global competitiveness and greater contribution to future waves of innovation. Furthermore, the S&T agenda of the future has to take into account changing global trends as well as the country's socio-economic needs. Existing scientific and technological capacity and growth opportunities may serve as a "beachhead" for integrating with global value chains, occupying an appropriate position in high-tech markets, and fundamentally improving the profitability of traditional sectors of the economy.

Acknowledgement The contribution was prepared within the framework of the Basic Research Programme at the National Research University Higher School of Economics (HSE) and supported within the framework of the subsidy granted to the HSE by the Government of the Russian Federation for the implementation of the Global Competitiveness Programme.

References

Bhagwati J (2004) In defense of globalization. Oxford University Press, Oxford

Cagnin C, Amanatidou E, Keenan M (2012) Orienting European innovation systems towards grand challenges and the roles that FTA can play. Sci Public Policy 39(2):140–152

Chulok A (2009) Forecast of S&T development prospects of the key economy sectors in Russia: future tasks. Foresight-Russia 3(3):30–36 (in Russian)

European Commission (2010a) European forward looking activities. EU research in foresightand forecast. Publications Office of the European Union, Luxembourg

European Commission (2010b) Facing the future: time for the EU to meet global challenges. Publications Office of the European Union, Luxembourg

Foresight 2030 (2014) Foresight for science and technology development of the Russian Federation for the Period up to 2030. Approved by the Prime Minister of the Russian Federation (No DM-P8-5 of January 3, 2014)

Georghiou L, Cassingena Harper J, Keenan M, Miles I, Popper R (eds) (2008) The handbook of technology foresight: concepts and practice. Edward Elgar, Cheltenham

Gokhberg L, Kuznetsova T (2011a) Strategy 2020: new outlines of Russian innovation policy. Foresight-Russia 5(4):8–30 (in Russian)

Gokhberg L, Kuznetsova T (2011b) S&T and innovation in Russia: key challenges of the post-crisis period. J East-West Bus 17(2–3):73–89

Gokhberg L (ed) (2016) Russia 2030: science and technology foresight. Ministry of Education and Science of the Russian Federation, National Research University Higher School of Economics, Moscow

HSE (2013) Long-term priorities of applied science in Russia. HSE, Moscow

KISTEP (2012) 10 future technologies. KISTEP, Seoul

Makarova E, Sokolova A (2014) Foresight evaluation: lessons from project management. Foresight 16(1):75–91

McKinsey Global Institute (2013) Disruptive technologies: advances that will transform life, business, and the global economy. McKinsey Global Institute, Chicago. http://www.mckinsey.com/insights/business_technology/disruptive_technologies. Accessed 6 Aug 2015

Meissner D (2012) Results and impact of national Foresight-studies. Futures 44:905–913

Meissner D (2014) Approaches for developing national STI strategies. STI Policy Rev 5(1):34–56

Meissner D, Gokhberg L, Sokolov A (eds) (2013) Science, technology and innovation policy for the future potentials and limits of foresight studies. Springer, Heidelberg, New York, Dordrecht, London

Message from the President of the Russian Federation to the Federal Assembly (2012) Official site of the President's office of the Russian Federation. On 12 December. http://kremlin.ru/news/17118. Accessed 6 Aug 2015

NISTEP (2010a) Future scenarios opened up by science and technology (summary). NISTEP report № 141. NISTEP, Tokyo. http://data.nistep.go.jp/dspace/bitstream/11035/676/1/NISTEP-NR141-SummaryE.pdf. Accessed 6 Aug 2015

NISTEP (2010b) The 9th science and technology foresight–contribution of science and technology to future society. NISTEP report № 140. NISTEP, Tokyo

OECD (2012) Looking to 2060: long-term global growth prospects. OECD Economic Policy Papers 3

Perez C (2002) Technological revolutions and financial capital. Dynamics of bubbles and golden ages. Edward Elgar, London

RAND (2006) The global technology revolution 2020: in-depth analysis. Technical report. RAND Corporation, Santa Monica

Sokolov A (2009) Future of S&T: Delphi survey results. Foresight-Russia 3(3):40–58 (in Russian)

Sokolov A, Chulok A (2012) Russian science and technology foresight—2030: key features and first results. Foresight-Russia 6(1):12–25 (in Russian)

Vishnevskiy K, Karasev O, Meissner D (2015) Integrated roadmaps and corporate Foresight as tools of innovation management: the case of Russian companies. Technol Forecast Soc Chang 90:433–443

Building a National System of Technology Foresight in Korea

10

Moonjung Choi and Han-Lim Choi

10.1 Outline of Korean Technology Foresight

Since the implementation of the first technology foresight (TF) in 1993–1994, TF in Korea has continuously advanced in response to society's increasing demands. The Framework Act on Science and Technology (S&T) in 2001, which specified regularly carrying out TFs, national TFs have been conducted every 5 years. In 2007, the third TF was revised to increase complementarities with the S&T Basic Plan, the nation's top-level plan in the field of S&T. The results of the revised TF were directly reflected in the second S&T Basic Plan. Furthermore, the results of the fourth TF (conducted during 2010–2011) found expression in the third S&T Basic Plan. All four TFs performed to date have primarily used the Delphi method. Since the third TF, future social trends were first identified and then future technologies predicted based on these trends; moreover, scenarios were developed founded on the results of the TF (Fig. 10.1). Currently, the Ministry of Science, Information and Communication Technologies (ICT) and Future Planning (MSIP) is responsible for TFs while the Korea Institute of S&T Evaluation and Planning (KISTEP) conducts the TFs.

The first TF aimed to identify a long-term development strategy for S&T. At the same time, Korea launched a large-scale, inter-ministerial R&D project (1992) which aimed at "raising the level of Korean S&T in the 2000s to the level of the

M. Choi (✉)
Office of Strategic Foresight, Korea Institute of S&T Evaluation and Planning, 12F Dongwon Industry Bldg., 68 Mabang-ro, Seocho-gu, Seoul 137-717, South Korea
e-mail: mjchoi@kistep.re.kr

H.-L. Choi
Creative Economy Innovation Center, Office of S&T Policy & Planning, Korea Institute of S&T Evaluation and Planning, 5F Trust Tower, 60 Mabang-ro, Seocho-gu, Seoul 137-717, South Korea
e-mail: airman10@kistep.re.kr

© Springer International Publishing Switzerland 2016
L. Gokhberg et al. (eds.), *Deploying Foresight for Policy and Strategy Makers*,
Science, Technology and Innovation Studies, DOI 10.1007/978-3-319-25628-3_10

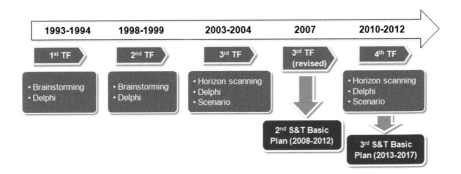

Fig. 10.1 Outline of Korean technology foresight. *Source*: compiled by authors

G7 countries". In 1993, Korea's national R&D budget exceeded one trillion won for the first time. In the first TF, S&T professionals determined 1174 future technologies over the next 20 years (1995–2015). Using the Delphi method, this TF surveyed the importance of future technologies, as well as their implementation time and technological level. In addition, the TF identified the factors hindering the creation of future technologies and the main actors in the development of future technology (Shin 1998).

The year 1999 saw the release of the results of the second TF. A Ministry for S&T had been created in 1998, as well as a National S&T Council set up in 1999. The purpose of the second TF was to study the future developments of S&T and to compare Korea's level of technology with that of more developed countries. This would enable policy makers to set goals for S&T policies and acquire the data needed for preparing a S&T strategy. In other words, the goal of the TF was to present a portfolio for the distribution of S&T resources nationally and to establish strategies for R&D projects based on the results of the TF. The second TF categorized the overall field of S&T into 15 areas, set the forecast period at 25 years from 2000 to 2025, and identified 1155 future technologies. As in the first TF, the 1999 TF employed the Delphi method to examine the importance, implementation time, and technological levels of future technologies. The 1999 TF also identified the main actors and the necessary policy measures for implementing future technologies (Lim 2001).

Figure 10.2 below indicates the conceptual diagram of the third TF, conducted from 2003 to 2004. Unlike the previous two TFs, the third TF considered the relationship between technology and society. In addition, the scope of participation expanded from S&T experts to include policy makers and social scientists. The third TF had three stages. The first stage identified the future issues and needs of society and the future technologies to address these needs. To organize future society's needs systematically, the TF separated them into individual, social, national, and global needs. Table 10.1 shows examples of future needs associated with individuals. Eight specialized divisions in the field of S&T were configured to determine future technologies; the forecast covered the period from 2005 to 2030, and identified 761 future technologies. The second stage evaluated the impact of

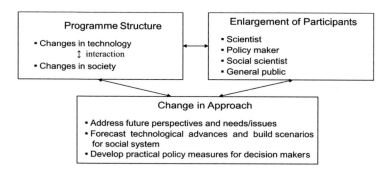

Fig. 10.2 Conceptual diagram of Korea's third TF (2003–2004). *Source*: Park and Son (2010)

Table 10.1 An example of future needs and issues in the third TF (individual level)

Actor	Need		Need or issue
	Main theme	Detailed theme	
Individual	Healthy life	Dealing with diseases	• Prevention, diagnosis and treatment of diseases that are hard to cure • Geriatric diseases • Chronic diseases • Contagious diseases • Artificial organs • Application of biotechnology
		Quality health service	• High quality healthcare system (ICT) • Alternative medicines • Secondary infection in hospitals
		Healthy normal life	• Comfortable daily life • Health-maintaining system
		Safe foods and consumer products	• Safer foods • Safer consumer products • Environmentally-friendly foods and consumer products

Source: Park and Son (2010)

various factors such as the implementation time of future technologies via the Delphi method. Finally, the third stage created scenarios about likely future challenges in education, health, labour resources, and security (Park and Son 2010).

How reliable have the previous TFs been in their predictions about future technologies? An evaluation of the 1109 future technologies predicted to exist by 2010 according to the first TF of 1994 found that 470 of these technologies were fully implemented and 331 partially implemented. This means the first TF's accuracy rate is 72.2 %, when we include the partially implemented technologies.

The partially implemented technologies include cases where the assessment of implementation depended on the viewpoint of the evaluation because any one future technology is determined by multiple technical factors or the concept of

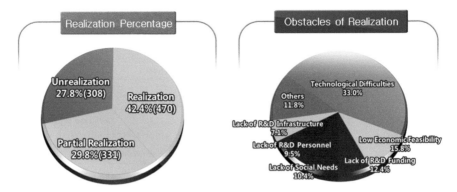

Fig. 10.3 Implementation percentages and obstacles to implementing the future technologies as identified by the first TF. *Source*: calculated by authors

the future technology is ambiguous. Using the Delphi method, the major obstacles for implementing future technologies were found to be, in descending order: "Technological Difficulties" (33.0 %), "Low Economic Feasibility" (15.8 %), "Lack of R&D Funding" (12.4 %), followed by "Lack of Social Needs" (10.4 %) (Fig. 10.3).

10.2 The Fourth Korean Technology Foresight

The fourth TF forecasts the future up to 2035 and had three stages (Fig. 10.4). The first stage forecast the future of Korean society and examined future needs. The second stage identified future technologies and conducted the Delphi survey to examine factors such as the technological implementation time and the time for socially penetrating future technologies, Korea's level of technology, the main actors for technological development, and governmental policies required for implementing technologies. The third and final stage created scenarios and illustrations depicting the shape of the future world that would be changed by implementing and distributing future technologies divided into 13 different areas, such as home and school. In addition, the fourth TF presented the possibility of various social changes caused by the development of future technologies by drawing up scenarios on the negative impacts of technological development.

10.2.1 Forecast for the Future Society and Discovering Future Needs

The fourth TF identified the most significant global trends that will affect societies up to 2035, or "megatrends". These megatrends are: (1) further globalization; (2) increasing conflicts; (3) demographic changes; (4) greater cultural diversity;

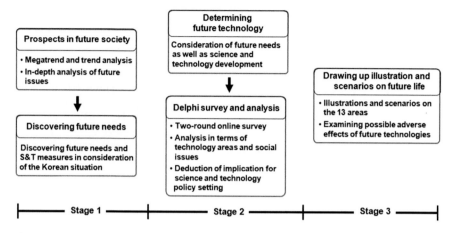

Fig. 10.4 Procedure of the fourth TF. *Source*: Choi et al. (2014)

(5) depletion of energy and resources; (6) greater climate changes and associated environmental problems; and (7) development and convergence of S&T. Furthermore, the continuing rise of China can also be considered a megatrend in its own right, which further accelerates the seven other megatrends. Table 10.2 below shows these eight megatrends, along with the 25 trends comprising the eight megatrends. For each trend, the TF examined the risks and opportunities for Korean society. Based on this analysis, the TF drew up the future society's needs.

10.2.2 Determining Future Technologies

The fourth TF defined future technologies as the "technology that can be implemented technologically or distributed socially by 2035 and has the potential for significant impacts on S&T, society, or economy." (NSTC and KISTEP 2012b). S&T experts determined future technologies in two ways, as shown in Fig. 10.5. One way is the *Demand Pull* type future technology, where future technologies capable of addressing the needs of future society are determined through forecasting future society. Another method is the *Technology Push* type future technology, where future technologies are expected to emerge from the development of S&T regardless of social needs. Technology Push type future technologies include technologies expected to emerge due to the accumulation of S&T knowledge, as well as technologies that currently only exist in conceptual form yet will become visible in the future. The fourth TF used methods such as patent trends, analyses of scientific papers, and technology roadmaps to determine future technologies.

This method enabled a list to be compiled of the 652 future technologies expected to emerge by 2035. As shown in Table 10.3 below, 601 (92.2 %) of the 652 future technologies are expected to emerge to address the needs of society in

Table 10.2 Megatrends and trends of the fourth TF

Megatrend	Trend	Megatrend	Trend
Further globalization	• Integration of the global market • Multi-polar world order	Greater cultural diversity	• More cultural exchanges and multicultural socialization • Improvements in women's status
	• Globalization of workforce • Extension and diversification of the governance concept • Rapid spread of epidemics	Depletion of energy and resources	• Increased demand for energy and resources • More shortages of food and water • Greater use of energy and natural resources for weapons
Increasing conflicts	• Deepening of conflicts between peoples, religions, and nations • Increase in cyber terrorism • Increase in risks of terrorism	Greater climate changes and environmental problems	• Greater global warming and increases in abnormal weather phenomena • More environmental pollution • Changes in ecosystems
	• Greater polarization	Continuing rise of China	• Increase in China's economic influence • Increase in China's diplomatic and cultural influences
Demographic changes	• Continuously low birth rates and ageing populations • Increase in urban population globally • Changes in the concept of family	Development and convergence of science and technology	• Development of information technology • Development of life science technology • Development of nanotechnology

Source: Choi et al. (2014)

the future, while only 51 technologies (7.8 %) are expected to appear because of developments in S&T. In addition, 394 technologies are related to more than two future trends, which means that more than 60 % of future technologies will address future needs related to plural trends. Examining the distribution of future technologies by sector reveals that there are over 90 technologies related to each of the following sectors: machinery, manufacturing, aerospace and astronomy, agriculture, forestry and fishery, and materials and chemical engineering. In contrast, the fields of information, electronics, and communication had the lowest number of technologies at 55 each (Fig. 10.6). The reason for the low number of

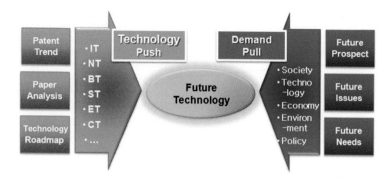

Fig. 10.5 Methods to identify future technologies used in Korea's fourth TF. *Source*: Choi et al. (2014)

Table 10.3 Future technologies identified by Korea's fourth TF and the distribution correlation between future trends

	Number of future trends related to each technology						Total
	Technology push	Demand pull					
Number of technologies	51	207	262	99	29	4	652
Proportion (%)	7.8	31.7	40.2	15.2	4.4	0.6	100.0

Source: compiled by authors

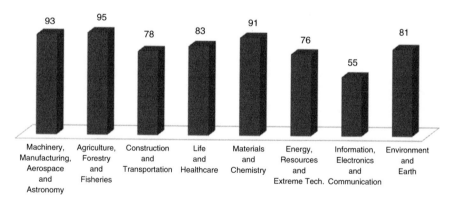

Fig. 10.6 Future technology distribution according to Korea's fourth TF. *Source*: calculated by authors

technologies in the latter fields is that technologies utilizing ICT, such as biosensor technology, are included in their field of application rather than in the ICT field. Table 10.4 below gives examples of detailed fields included within their respective broader areas.

Table 10.4 Examples of the detailed fields within each field of the fourth TF

Field	Detailed field
Machinery, manufacturing, aerospace and astronomy	Manufacturing and process, robot, space and exploration, satellite, aircraft, unmanned aerial vehicle, automobile, shipbuilding, defense, counterterrorism, etc.
Agriculture, forestry and fishery	Crop production, animal science, animal disease, zoonoses, fish farming, tree breeding, forest environment, customized food, etc.
Construction and transportation	Construction material and equipment, building control management system, railroad, aviation, distribution, safety management, etc.
Life and healthcare	Brain science, pathogen measurement, medical engineering, cancer diagnosis and treatment, medicine, artificial organ, oriental medicine, etc.
Materials and chemical engineering	Functional alloy material, nano-sensor, semiconductor material, medical material, battery, carbon nanotube, chemical process, etc.
Energy, resources and extreme technology	Smart grid, electric power, nuclear energy, resource and exploration, solar energy, extreme technology, etc.
Information, electronics and communication	Virtual reality and augmented reality, display, sensor, telecommunication, information protection, information theory, etc.
Environment and Earth	Weather and climate, air quality management, ecosystem restoration, carbon capture and storage, eco-friendly material and process, earthquake, marine environment, etc.

Source: compiled by authors

10.2.3 Delphi Survey and Analysis

The fourth TF included a two-round online Delphi survey, which twice collected the opinions of experts. Responses were received from 6248 people in the first round and from 5450 in the second round. The number of respondents who participated in the first and second rounds increased significantly compared to the first three TFs (Table 10.5). Table 10.6 summarizes the Delphi survey questions used in the fourth TF.

The average time for technological implementation of future technologies as determined by the fourth TF was by 2021. The TF predicted their widespread distribution across society to occur in 2023.

When comparing each field's results with the average technological implementation time estimates, the field of information, electronics and communication was the fastest (year 2019), while the field of life and healthcare was the slowest (2022 year). Experts surveyed agreed that 519 future technologies (79.6 % of the total identified) will be technologically implemented in Korea within the next 10 years (by 2022). Furthermore, they predicted that 294 technologies would be distributed across society within the same period of time. 2.7 years is the predicted average time for future technologies to be widely implemented across society after technological implementation.

Table 10.5 Number of Delphi survey respondents in four TFs

		1st TF	2nd TF	3rd TF	4th TF
Number of future technologies		1174	1155	761	652
Response	1st round	1590	1833	5414	6248
	2nd round	1198	1444	3322	5450

Source: calculated by authors

Table 10.6 Delphi survey items of the fourth TF

Survey item	Survey content
Technology level	Nation at the forefront of the technology level
	Korea's technology level
Technological implementation time and social distribution time	Implementation time, and general public use time in Korea
	Realization time and general public use time in technologically most-advanced nations
Technological implementation measures	Main actors in R&D
	The need for collaborative research
Role of government	The need for government investment
	Government priority measures to be implemented
Importance in future society	– Contribution with respect to technology aspects – Contribution with respect to public benefits – Contribution with respect to economy and industry
Possibility of negative effect	Possibility of negative effect caused by general public use
Institutions involved in research	Local and international research institutions
Interrelationship with future trends	Relationship to 22 future trends[a]

[a]Three technology-related trends were excluded from the 25 future trends
Source: compiled by authors

When examining the current state of countries with the highest level of technology in relation to the 652 future technologies, the fourth TF found that the United States possesses the highest level of technology in 495 technologies. Japan was second with 141 technologies, and the EU was a distant third with 32 technologies. The research revealed that Korea's average technology levels were 63.4 % of the leading countries regarding the 652 future technologies. The level of technology for 18 future technologies was above 80 %, which indicates that Korea leads the field in these technological areas, with nine included in the field of information, electronics and communication, which is more than that in any other field. At the same time, the study found that the levels of 22 technologies were below 40 % and thus were part of the "lagging" group, among which nine were in the field of machinery, manufacturing, aerospace and astronomy. Of the 652 future technologies, Korea's highest technology level was in "terabit level next-generation memory device technologies" (90 %).

Table 10.7 Priority policies to support the development of future technologies by time needed for technological implementation (based on the results of Korea's fourth TF; share of respondents who chose each option, %)

Technological implementation time	Increased R&D funding (%)	Greater collaboration (%)	R&D personnel training (%)	Infrastructure construction (%)	System improvements (%)
Short (–2017)	28.5	20.0	16.9	22.1	12.5
Medium (2018–2022)	31.7	22.9	18.9	18.9	7.6
Long (2023–)	31.9	23.7	22.3	18.3	3.8
Total	31.6	22.8	19.5	19.0	7.1

Source: calculated by authors

Table 10.8 Examples of future technologies with high likely negative impacts

Technology for building underground waste storage
Personal life log technology which can make database by saving personal life with sound and image data
Gene therapy technology for fetus
Electro Magnetic Pulse (EMP) bomb disturbing electronic parts in the enemy's weapon system by detonating it in the air of the enemy
Technology of developing functional transgenic fish species that can produce useful substance (nutritional contents, medicine and medical supplies)
Conversion technology from uranium-238 to plutonium-239 using the liquid metal reactor

Source: compiled by authors

When examining the priority policy measures that the government should enact to help implement future technologies, the survey results showed that most respondents felt increased R&D funding was highest priority (31.6 %). The next most important policies stated were greater collaboration, training for R&D personnel, and infrastructure construction. The need for system improvement ranked the lowest; however, it was higher in the construction and transport field (13.8 %) and environment and earth sciences (10.4 %) than other fields. Future technologies with faster implementation times placed more value on infrastructure construction and system improvements, while future technologies with longer implementation times attributed relatively more value to R&D personnel training and greater collaboration (Table 10.7).

Unlike the previous TFs, Korea's fourth TF asked about any unintentional negative impacts on society, culture, or the environment of the widespread social distribution of future technologies. Table 10.8 lists some examples of future technologies with relatively high likelihoods of negative impacts. Six of these technologies were included in a future scenario, accompanied by an analysis of both the positive and negative effects.

10.2.4 The World of the Future Changing Through S&T

The main purpose of TF in Korea is to predict the development of S&T and use the results in developing S&T policies. However, informing the public about what the future holds based on the development of S&T is an equally important role of TF. For this purpose, the future world is divided into 13 different areas (home, school, hospital, office, factory and plant, transportation, fishing village, farming village, city, disaster, space, war and terror, and underground) and each area is

Fig. 10.7 Illustration and scenario of a future society (a family in 2035).
The phone rings while Jung-Hoon and his wife are watching TV. Their daughter appears, smiling brightly, on the TV screen. For a second, Jung-Hoon and his wife think that the drug which their daughter has been taking for three months for depression caused by her inability to become pregnant is effective. The drug their daughter takes is a **non-addictive** chemical that she can take any time **to enhance positive emotions, such as happiness, without causing harm to her body**. The drug regulates crime-related emotions in a calculated manner and improves brain capacities, such as reasoning skills, creativity, and memory storage abilities; therefore, it is used in rehabilitating criminal offenders and rehabilitation education as well as a supplement food for students preparing for a test.
The daughter's news that they hear as soon as they answer the phone brings joy greater than her bright face. "Dad, Mom, I'm pregnant!" The daughter says that the device she received a while back from her friend greatly helped, and she begins to explain the process one by one, from the strange feelings she was getting these days to visiting the hospital and hearing the news of her pregnancy. The friend's present is **a portable device that tells the user her biological cycle, and it diagnoses the bio-molecular changes related to pregnancy and predicts the possibility of pregnancy to inform her when she is at her optimal fertile period**.
After the phone call with her daughter, Mi-Young runs into the room. "What are you up to?" asks Jung-Hoon as Mi-Young opens the drawer looking for something. "We must stay healthy if we want to see our grandchild... Found it!" Mi-Young heads to the bathroom holding a cancer diagnostic kit. **The cancer diagnostic kit is a self-diagnostic kit that is able to identify the five major cancers from urine and even helps the user discover cancers with very few early symptoms, such as liver cancer.** "What are you doing? You should also get tested." As Jung-Hoon holds the diagnostic kit waiting for the results, a smile spreads across his face as he thinks about the new family member they will soon meet.
Source: NSTC and KISTEP (2012b)

connected with the future technologies. The scenarios and accompanying illustrations are composed by classifying the future into 10 years later (year 2022) and the year 2025 to compare the future world over time (Fig. 10.7). The periods of time when the future technologies are likely to be widely distributed—as determined through the Delphi survey—were used as the reference points for selecting the future technologies specific to each point in time.

10.2.5 Policy Implications

By analysing the Delphi survey results about future technologies as part of Korea's fourth TF, we identified the following policy implications.

First, among the share of Korea's technologies belonging to the leading group (level of over 80 %) or the next group (level of 61–80 %), 344 future technologies were expected to be implemented within the next 5–10 years. Korea has the possibility of joining the world-leading group if it pursues R&D more actively. However, Korean technologies are not yet at the highest level. Therefore, policy support for developing unique technologies is necessary. It is important to develop a diversity of 'technology sprouts' since future technologies since success remain largely uncertain.

Second, achieving advanced technologies effectively is possible through joint R&D efforts of industry, universities, and research institutes as well as through international collaborative research. Furthermore, the high demand for R&D personnel training and infrastructure construction, as identified by our analyses, indicates that future technologies require systemic support in the medium and long-term. At the same time, it is important to implement policies to minimize the adverse effects of certain future technologies through Technology Assessment mechanisms.

Third, it is important to pay sufficient attention to the social aspects of S&T, given the importance and possible consequences (the ripple effect) of social problems in the future. By carrying out national R&D projects that address these kinds of issues, the effectiveness of S&T in responding to future issues will be strengthened (together with other solutions). To achieve this goal, it is necessary to analyse future issues and existing factors, including the implementation time of future technologies, the level of technology, and the R&D plan. Furthermore, an optimal technological development strategy is required that would set out the priorities taking into account the roles of each technology in solving specific issues, and clarify the roles and accountability mechanisms of various government departments and research institutes.

10.3   Technology Foresight and S&T Planning

Korea develops a new S&T Basic Plan every 5 years alongside the launch of a new administration. All S&T planning activities at the national level are connected with the S&T Basic Plan (Fig. 10.8). The National Mid- & Long Term S&T Development Strategy selects the national core technologies based on the future

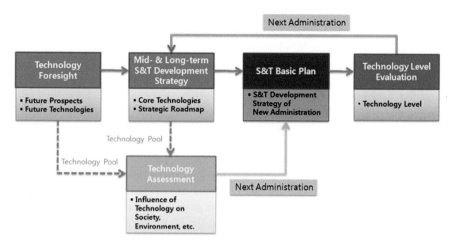

Fig. 10.8 S&T planning activities in Korea. *Source*: compiled by authors

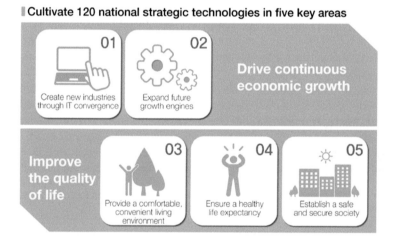

Fig. 10.9 The third S&T Basic Plan: developing national strategic technologies. *Source*: MSIP and KISTEP (2013c)

technologies determined by the TF and establishes a strategic roadmap for these technologies. These results are reflected in the focused initiatives related to the technological development of the S&T Basic Plan. The 652 future technologies identified by the fourth TF went through a reviewing process by committees responsible for the national R&D budget as well as by R&D-related ministries. As a result, a list of 120 national strategic technologies was compiled; these technologies were also identified by the third S&T Basic Plan (Fig. 10.9).

The importance attributed to Korea's TFs has grown with every round. Accordingly, the third TF—in contrast to the two preceding TFs in which only scientists and engineers participated—examined future social development and factors of demand to identify future technologies that could address society's demands. Moreover, the third TF was modified to ensure closer integration with the 5-year S&T Basic Plan and help provide a systemic basis for national level S&T planning. In this process, TFs have consistently provided background information that feeds into policies and medium and long-term S&T strategies.

In addition to the TF, Korea regularly conducts Technology Level Evaluation and the Technology Assessment exercises (Fig. 10.8). The Technology Level Evaluation targets the national strategic technologies as indicated in the S&T Basic Plan, and takes place every two years. The Technology Level Evaluation exercise compares the technological levels of Korea, the United States, China, Japan, and the EU using Delphi survey methods, patent analyses, and research paper analyses (MSIP and KISTEP 2013b). Those who devise strategic roadmaps for national core technologies use the results of this evaluation as inputs.

The Technology Assessment evaluates the positive and negative impacts caused by new S&T advancements on areas such as the economy, society, culture, ethics, and the environment. It also suggests ways to enhance the positive impacts and avoid the adverse effects. Korea conducts a Technology Assessment annually, and as part of this assessment surveys not only experts from the humanities, social sciences, and S&T, but also members of the public. Recently, the country undertook a Technology Assessment on big data (NSTC and KISTEP 2012a) and 3D printing (MSIP and KISTEP 2013a). The future technologies identified by TFs were used in the process of selecting a target technology for Technology Assessment. The results of the Technology Assessment are reflected in the research plans regarding R&D projects in the corresponding fields. Furthermore, the results of the Technology Assessments are not only taken into account when formulating the S&T Basic Plan but also when devising policies to minimize the negative impacts of new technologies.

Acknowledgement   The paper is a revised and developed version of a previously published paper by Choi M., Choi H. (2015). Foresight for Science and Technology Priority Setting in Korea. Foresight and STI Governance, vol. 9, no 3, pp. 54–67 with permission from Foresight and STI Governance journal (http://foresight-journal.hse.ru/en/).

References

Choi M, Choi H, Yang H (2014) Procedural characteristics of the 4th Korean technology foresight. Foresight 16(3):198–209

Lim K (2001) The 2nd technology forecast survey by Delphi approach in Korea. NISTEP Res Mater 77:105–118

Ministry of Science, ICT & Future Planning and Korea Institute of S&T Evaluation and Planning (2013a) Technology assessment: 3D printing (in Korean)

Ministry of Science, ICT & Future Planning and Korea Institute of S&T Evaluation and Planning (2013b) Technology level evaluation: 120 national strategic technologies of the third S&T Basic Plan (in Korean)

Ministry of Science, ICT & Future Planning and Korea Institute of S&T Evaluation and Planning (2013c) The third science and technology basic plan

National Science & Technology Commission and Korea Institute of S&T Evaluation and Planning (2012a) Technology assessment: Big data (in Korean)

National Science & Technology Commission and Korea Institute of S&T Evaluation and Planning (2012b) The fourth technology foresight (in Korean)

Park B, Son S (2010) Korean technology foresight for national S&T planning. Int J Foresight Innov Policy 6(1/2/3):166–181

Shin T (1998) Using Delphi for a long-range technology forecasting, and assessing directions of future R&D activities: The Korean exercise. Technol Forecast Soc Chang 58(1/2):125–154

Foresight, Competitive Intelligence and Business Analytics for Developing and Running Better Programmes

11

Jonathan Calof, Gregory Richards, and Jack Smith

11.1 Introduction

Establishing industrial policy and its ensuing programmes and industry' assistance measures is fraught with high levels of uncertainty. As will be shown in this article, an integrated programme involving foresight, intelligence and business analytics not only will decrease levels of uncertainty and risk, but these techniques should lead to increasing probabilities of policy uptake by its intended audience and also early identification of industry opportunities.

This article is based on both academic scholarship and practitioner experience. The authors of this article have been involved collectively in hundreds of industrial policy and programme projects around the world and in many academic studies. Dr. Jonathan Calof has approached programmes and policies through the field of competitive intelligence, Dr. Greg Richards through business analytics and Jack Smith through foresight. In this article, each of these disciplines will be defined and it will be shown using an example how and why the three approaches can be combined, as well a dashboard applying these concepts will also be demonstrated.

To provide a common base for discussing the three domains in the context of programme development and monitoring the following fictitious example is used. The government has noted that demand for and talk about nutraceutical products is growing and this could represent an enormous opportunity for Canada. Accordingly, the government wishes to design a programme that will encourage Canadian companies to produce innovative nutraceutical products and technologies. It is hoped that this will stimulate products, which will become commercialized, leading to jobs in the sector and wealth creation. Similar to other programmes the

J. Calof (✉) • G. Richards • J. Smith
Telfer School of Management, University of Ottawa, 55 Laurier Avenue East, Ottawa, ON K1N6N5, Canada
e-mail: calof@telfer.uottawa.ca; richards@telfer.uottawa.ca; jesmith@telfer.uottawa.ca

© Springer International Publishing Switzerland 2016
L. Gokhberg et al. (eds.), *Deploying Foresight for Policy and Strategy Makers*,
Science, Technology and Innovation Studies, DOI 10.1007/978-3-319-25628-3_11

programme is envisioned to be a tax credit for eligible nutraceutical R&D and commercialization investments.

11.2 Foresight

Foresight involves constructively bringing awareness of long-term challenges and opportunities into more immediate decision-making. "Foresight is a systematic, participatory, future-intelligence-gathering and medium-to long-term vision-building process aimed at present-day decisions and mobilising joint anticipatory-preparatory actions" (For-Learn 2014).

Foresight is neither prediction, nor does it estimate probabilities of particular pathways. Rather, it is about broadening our understanding about the drivers of societal change and becoming better prepared for the inevitable surprises ahead. Foresight normally starts with scanning to determine what is changing and why by anticipating plausible sources and origins of change, and seeking to understand the multiple complex interdependencies that motivate personal adaptation, organizational positioning—capacity for adjustment, and—at a more macro level—societal evolution. Foresight then uses the various change prospects to construct a range of plausible narratives, or scenarios, and roadmaps to indicate basic directions.

Foresight asks *"what range of plausible futures might our organization have to be prepared for, and which strategies can help us build resilience and create adaptive capacity to anticipate and thrive in the turbulence of change?"*

Foresight increases organizational agility through added resilience—alertness to trends, awareness of change drivers and readiness for potential shocks, issues, and challenges Essentially, foresight employs a rehearsal approach to preparedness by addressing *what if* scenarios.

Foresight establishes a context (i.e. boundaries and possibilities of what are deemed plausible narratives) for both the extent and speed of potential change and the adaptive risks of a designated sector, emerging market or technology domain.

11.3 Applying Foresight to Nutraceuticals

So how might the government apply foresight to nutraceuticals? How could foresight be used to address concerns about the long term competitive viability of its nutraceutical sector and approach the challenges of a rapidly changing technology landscape? More specifically, how should an Industry department, already with a successful track record of nurturing the development of new nutraceutical companies approach the complexity of deciding whether, when and how to invest in a new area of technological progress with potentially transformative applications?

Table 11.1 Results of a foresight STEEP exercise by Saritas and Smith (2008)

Society and culture	Social norms, education, information and knowledge society	Demographics, urbanization, population health and migration	Equity, ethical, moral and legal issues
Science and technology	Science culture and discoveries	Technology progress	Innovative, transformative, applications and products
Energy	Current energy use, peak oil, efficiency and security	New and renewable resources	Non-renewable energy alternatives
Ecology-Economy	Stage of global finance, trade, debt and related globalization issues	BRIC rapid development economies	Climate change, global warming = sustainable ecology, new economy

A first step would be to consult technology and environmental scanning reports similar to those categories highlighted in Table 11.1.[1] The survey used five standard STEEP-type categories (four presented in the table) and then three sub-categories for each—with technology areas embedded in each sub area, and/or featured in the two highlighted areas under Science and Technology).

This foresight technique indicates that there exist real uncertainties about what applications might soon become both technically feasible and economically viable and whether there may be toxicological risks. Applying this within nutraceuticals found that there was already a growing global market in nutraceuticals—and—that the application of molecular scale nano-engineering was progressing fast and could create enormous growth if and when successfully commercialized. Significant uncertainty however remains around which countries and producers could do this when and how.

To better understand the broader context of these uncertainties two additional foresight techniques are frequently employed: scenarios and technology roadmaps [see Smith and Saritas (2011) and Popper (2008) for more on foresight analytical technique and method for selection].

Scenarios explicitly build upon identified key uncertainties to develop future oriented situational narrative visions and glimpses of plausible future operating environments that can reveal business challenges as well as opportunities flowing from the resolution of the identified uncertainties—thereby enabling anticipatory actions in advance of one's competition.

For example in the area of future nutraceutical applications, four representative scenarios could be derived from, for example, the dual uncertainties of science and technology (S&T) rate of progress, and the pace and performance results of regulatory oversight (these drivers are based on past scenario projects in similar

---

[1] Table 11.1 was developed by Jack Smith and Dr. Saritas as a contribution to the 2008 European Commission's Future Technologies Assessment Conference overview report on the Big Picture Survey.

areas that the authors have been involved with). In this example, four different scenarios emerged that were called nutri-slow, nano-go, nutri sue, and nano promo. Note that two of the scenarios involved nanotechnology and what was most apparent was that regardless of whether oversight was uncertain or high, as soon as the dynamic for S&T progress shifts into fast—the result moves into the nano driven zone. To apply this to programme developed we first need to determine: where we are now (i.e. in 2014–2015); where we seem to be heading; and whether this can or should be changed in some manner through policy actions. So what messages are the foresight scenarios conveying?

- The current market—2014—for conventional nutraceuticals is projected to remain sluggish but could soon become highly vulnerable if (as expected by leading scientists), nano-scale design and production advances enable producers in other countries or markets to shift into what is described as fast and transformative—a situation that would create a new base for competition.
- There are understandable uncertainties associated with R&D, regulatory approval which will have to be closely monitored—because if the new nano techniques are able to obtain approval then current production platforms will become as obsolete as floppy disks competing against flash drives.
- While timelines are imprecise in foresight, it is clear that key change factors—as represented by the scenario drivers and uncertainties—are going to be influencing the next business cycles of nutraceuticals.

More specific to the needs of most business enterprises than scenarios (which are typically initiated by governments) technology roadmaps are employed to further reduce uncertainty; first by being managed by industry, and second, by having more near-term and specific decision time lines for investment—i.e. what specific investments will be required and when (e.g. new R&D; equipment, training and skills development, emerging market research) to acquire the needed agile capacity to realize the opportunities and to reach the business destination before others.

Further analysis into nano composite new materials leads us to the possibility of nano based nutraceuticals or nano-nutraceuticals, which would likely score in the high range with moderate risk in terms of policy barriers.

Although technology foresight is showing that several nano-nutriceuticals have already been commercialized, risk nevertheless remains, mainly because the regulatory environment has not yet fully rendered its judgments and concerns about health implications around ingesting nano based products.

A typical foresight insight or conclusion of the technology roadmaps application is that:

- The matrix analytical framework suggests positive potential from the new technological opportunities.
- Further R&D will be required, especially in terms of the regulatory hurdles.
- To succeed—or at least to be early entrants into the emerging new nano-based nutraceutical design and production platforms will require excellent scientific

capabilities and equipment that aspiring firms must plan and recruit for in advance if they want to be competitive.

In conclusion by use of foresight, a programme recommendation of nano-nutraceuticals has been made. This has arisen through STEEP and scenarios. A nano-nutraceutical roadmap has provided the information needed to focus the programme on specific kinds of research and issues such as regulations to address.

11.3.1 Competitive Intelligence

Definitions of competitive intelligence focus either on the objective of CI or how CI is done (process definition). For example, the Strategic and Competitive Intelligence Professionals (SCIP), a global association of competitive intelligence practitioners, defines competitive intelligence as "CI is a necessary ethical discipline for decision making based on understanding the competitive environment" (SCIP 2014). While this does not define competitive intelligence it does describe its objective. Similarly Professor Du Toit defined intelligence in terms of its objective "Competitive intelligence (CI) is a strategic tool to facilitate the identification of potential opportunities and threats" (Du Toit 2013). Salvador et al. (2013) wrote that the objective of intelligence was to support innovation.

Others have defined intelligence in terms of its process, i.e., how it is created. For example, Kahaner (1997, p. 16) wrote that CI is "a systematic programme to collect and analyze information about competitors' activities and general business trends to achieve the goals of the company". Moreover, "CI comprises identification of intelligence needs within an organization, collection of data from primary and secondary sources, evaluation and analysis". Kahaner's definition fits with the concept of the wheel of intelligence which posits that intelligence is developed in a systematic and ethical manner involving planning, collection, analysis, communication and management.

The field has a very long and rich academic and practitioner history with academic literature cites appearing in the 1950s and company practices noted in the fifteenth and sixteenth century (Juhari and Stephens 2006), thus, it can hardly be called a new discipline. However, with the increasingly competitive environment, government and business have been turning more and more to competitive intelligence as a method for better understanding their environment and developing better programmes and strategies. A study done by Global Intelligence Alliance (GIA 2011) noted that the percentage of respondents with intelligence functions had grown from 63 to 76 % in 2 years and that one third of the companies that didn't have an intelligence operation yet intend to launch one within 12 months (GIA 2011). A study done by American Futures Group consulting firm found that 82 % of large enterprises and over 90 % of the Forbes top 500 global firms adopt CI for risk management and decisions (Xu et al. 2011). The Xu study also pointed to a high value of the CI industry with an estimate that by the end of the twentieth century, the overall production value of CI industry had reached US$70 billion

(Xu et al. 2011). A more conservative figure of $2 billion a year was provided by SCIP (SCIP 2014). Regardless of the figure used studies do report that the amount spent on competitive intelligence is growing and also that intelligence was paying off. A GIA study reported that decision making was 15 % more efficient in companies with an intelligence function in place and 80 % said the investment is paying off in terms of return on investment (GIA 2013).

In trying to understand competitive intelligence practice various organizations have surveyed intelligence practitioners. Global Intelligence Alliance (www. globalintelligence.com) does these studies on a regular basis and two were described above. Academics throughout world have looked at their country's CI practices, sometimes on a comparative basis (see for example Wright and Calof 2006; Du Toit 2013; Bergeron 2000). In 2005, the Competitive Intelligence Foundation supported a global study on competitive intelligence practice (Fehringer et al. 2006) among its findings was that intelligence was being used to help make many different kinds of decisions including market entry, product development, R&D, corporate etc. (Table 11.2). The study also pointed to a broad range of analytical methods used for developing intelligence (Table 11.3). In terms of where the information for developing intelligence came from, consistent with exiting intelligence theory, the number one source on primary information came from the organization's own employees, followed by industry experts and customers with conferences and trade show's being amongst the more dominant places for gathering information from primary sources. In terms of secondary sources, while Internet was highly used (85 % responded very important or important), online and print publications were higher at 97 % and fee based online subscription was similar to internet at 84 %.

Intelligence has several subdomains or speciality fields including competitor intelligence (intelligence focused on competition), sourcing intelligence (intelligence used in the human resource function) and most relevant for this article competitive technical intelligence (CTI). CTI is competitive intelligence within the R&D arena (Ashton and Klavans 1997). Ashton and Klavans defined it as 'business sensitive information on external scientific or technological threats,

Table 11.2 What business decisions do your departments CI decisions support (%)

Decision supported	Frequently	Sometimes	Rarely	Never	Don't know
Corporate/business decisions	54.1	32.7	8.5	3.2	1.6
Market entry decisions	38.9	38.3	13.6	5.7	3.5
M&A, due diligence joint venture	25.9	31.3	22.2	14.6	6.0
Product development	36.8	37.2	16.6	5.7	3.6
Regulatory or legal	12.9	30.7	30.5	17.4	8.6
Research or technology development	24.4	39.2	21.0	10.3	5.1
Sales or business development	48.7	35.8	10.3	2.4	2.8

Source: Fehringer et al. (2006)

opportunities, or developments that have the potential to affect a company's competitive situation' (Ashton and Klavans 1997, p. 11). Literature referencing CTI goes as far back as the 1960s. For a more detailed look at CTI see Calof and Smith (2010).

11.3.2 Government Use of Competitive Intelligence

While much of the intelligence literature focuses on the use of competitive intelligence by companies to support economic and technical decisions, there is a stream of literature that looks at its importance for governments. Growth in government use of competitive intelligence led to SCIP in 2004 allocating a conference track to government and CI. Driving the increased use of CI by the public sector is the difficult financial, economic and political decisions facing public managers and both the need for and availability of competitive intelligence techniques to help with these decisions (Parker 2000). Calof and Skinner (1999) looked at competitive intelligence within the Canadian government environment noting that it was being used extensively in various departments for policy development. At a technical intelligence level, Fruchet (2009) wrote that the CTI group at the National Research Council (Canadian government organization) "provided technology intelligence products and services to the business and market development customers in both the NRC research institutes and the Industrial Research Assistance Program" (Fruchet 2009, p. 37). The authors of this paper have worked with governments in many countries (and provinces in Canada) in developing intelligence programmes that have been used for stakeholder analysis, treaty negotiations, identification of international priorities, developing technology programmes and policy, and more.

Canada is not unique in terms of government using competitive intelligence processes to develop programmes and policies. Bonthous (1995) examined the French government's use of competitive intelligence for policy and programme development and Gilad (1998) looked at the Japanese model.

11.3.3 Foresight as a Complement to CI and CTI

Calof and Smith (2010) developed a framework for R&D project selection that combined foresight and competitive technical intelligence. In the article they describe the two as a complementary approach: "Today's decisions will shape the environments of tomorrow whether in business or government, and however one acquires the best intelligence, new market characteristics and estimates and a disciplined imagination of plausible situations, the agility of positioning and response can be substantially increased through a complementary approach that, if successful in capturing the many dimensions of future risk, will represent an integrated capability" (Calof and Smith 2010).

The perspective of foresight as was noted in the earlier section is outside in, that is it looks outside the frame of any organization or country strategy and asks the question what does the future environment look like? Intelligence on the other hand starts with an existing strategy and asks the question how will the environment impact the success of this strategy? Foresight tends to look long term (in some cases 50 years out), the time frame for competitive intelligence is considerably shorter. The timeline for CTI is generally longer than that of other forms of intelligence but far shorter than foresight. Calof and Smith (2010) reviewed several CTI studies finding them to be generally within the range of 3–10 years. Foresight as mentioned earlier broadens understanding and identifies pathways while intelligence takes those pathways and understanding and within the short to medium term looks for estimates, probabilities and forecasts to help companies adapt their strategies to the most likely environmental context. Intelligence adopts a predictive approach to scoping future risks that seeks to provide direction to decision makers on the implications of new and emerging technologies and their prospective markets. The expected outcome is more effective organizational development and competitive strategies compared to foresight which is broader. Together, foresight and competitive intelligence contributes collection methodologies including primary and secondary collection approaches, facilitation methods, a variety of robust analytical methods, ability to work with qualitative information and a clear focus on understanding the external environment, as such they are highly complementary.

11.3.4 Application of Intelligence for Design of the Nano-Nutraceutical Programme

First, foresight provided the decision maker with two very valuable inputs—rather than the generic concept of nutraceutical the scenario exercise focused it on nano-nutriceutical. Second, roadmapping identified some of the issues that will have to be addressed enroute to commercialization (e.g. regulatory), and company requirements (for example R&D capability).

How would a competitive intelligence professional continue with the programme development? What would their contribution be that would be different from the foresight contribution? As the previous section noted, intelligence will potentially adopt a shorter term orientation and the focus will be more on the strategy.

Foresight has recommended that the government develop a programme that will encourage Canadian companies to engage in appropriate nano-nutraceutical research and commercialization. If the objective is to get companies to do this kind of research, intelligence will ask two questions:

1. Are Canadian companies willing to engage in this kind of research (interest and capacity)?
2. What incentive will they require to engage in this kind of activity? For example, a loan guarantee and if what so at what percentage? Is it a tax credit? If so at what level? Grant programme?

In general the intent of this type of government research assistance programme is to get companies to change their R&D behavior to match the government's desires. How do you get a company to change its behaviors? There are many analytical techniques in intelligence [see Fleisher and Bensoussan (2002) for a description of some of the more popular ones] and for a question like this, intelligence turns to a technique called profiling which involves putting together a detailed psychological based assessment of the target. Profiling asks how will the target most likely react? In understanding the target, profiling also is able to answer the question, what do I have to do to get the designated reaction from the target. A competitive intelligence profiler would seek to develop detailed profiles on the companies that would be likely to adopt the government's research programme. The profiler would be looking for information about company's research decisions including what drives these decisions and the risk orientation of the target. The profile needs to be designed to determine both potential companies interest in doing nano-nutraceutical research and what kind of incentive would lead a company to make this decision. Most of the information required for this kind of profile the companies should be readily available to the government. For example, the companies being profiled may have already applied for programmes, associations may have presented reports and recommendations to the government, past programmes can be examined as well and a check into the programme files and discussions with programme officers who oversaw the programme would help.

Table 11.3 (intelligence analytical techniques) lays out many of the more popular competitive intelligence analytical techniques and most of these come under the category of strategic analysis and environmental analysis. The reason that environmental and strategic analysis techniques are so popular is they get at the

Table 11.3 How often do you or others in your department use the following analysis techniques?

Technique	% Used	Technique	% Used
Strategic analysis techniques		Competitive and customer	
BCG Matrix	46.2	Blindspot	54.3
Industry analysis (5 forces)	78.1	Competitor	90.1
Strategic groups	64.3	Customer value	74.2
SWOT	90.3	Customer segmentation	79.6
Value chain	65.6	Management profiling	70.5
Environmental analysis		Evolutionary analysis	
Issue analysis	69.1	Experience curve	48.8
Scenarios	68.6	Growth vector	47.0
Stakeholder	61.8	Product life cycle	68.2
STEEP	59.9	Technology life cycle	65.0
Financial analysis			
Financial ratio	76.1		
Sustainable growth rate	66.5		

Source: Fehringer et al. (2006)

heart of what decision makers need to know, is the market a profitable one and what do we as a company needs to do to capture those profits. Therefore an intelligence practitioner will want to do a market profile. Table 11.3 also has evolutionary techniques that look at technology direction within an industry. These are important because the intelligence officer wants to ensure that the kind of research that is being encouraged will be appropriate for the future environment. If it will take 5 years for companies to conduct the research and get something ready for commercialization then the intelligence practitioner will ask where is likely to be in the next 5 years? Where is the industry going? What can I expect competitors to do over the next 5 or more years (surely they will not be sitting on today's technology for 5 or more years).

Among the more popular techniques used for this purpose is timelining. Intelligence realized a long time ago that there were logical sequences to any major shift in a marketplace. For example, long before a new technology hits the market there had to have been manufacturing activity, before that testing, before that research and so forth. Each of these sequenced steps leaves information that those interested can view. For example research activities are accompanied with patents and sometimes poster sessions at conferences. It is no wonder that several companies have told employees that when they see someone new at a trade show to let management know. The new player could be a potential customer, competitor, supplier etc. learning about the industry.

Similarly techniques such as science mapping have been developed to look at what research communities are coming together to better project research direction. While projecting 10 years out is very difficult, timelining makes things a little more certain by looking at what activities have already been done. The idea then is to identify what is currently happening in the industry at a global level and place it on the timeline. The information is taken from secondary sources such as magazines and various online databases but is also more commonly found by attending industry events. Event intelligence is a growing discipline within intelligence and involves organizing for data collection at conferences, tradeshows, workshops and the like to gather this kind of information. By use of event intelligence it should be relatively straightforward to identify how fast and transformative science currently is, and where various companies are positioned on the timeline. Take for example the following quote from Forbes magazine:

> Nestle may also be exploring nutraceuticals–nano-capsules that deliver nutrients and antioxidants to specific parts of the body at specific times. The technology turns previously insoluble nutrients into nano-sized particles that can be released into the body and properly absorbed, with big potential benefits for a whole new kind of health food (Wolfe 2005).

Clearly Nestle's is well along the development curve. If this is where they potentially were in 2005, the intelligence officer will project, through timelining, where they likely to be in 2014 and based on this the likely state of their research and commercial offerings in 2020 (the targeted commercialization period envisioned by the program). Whatever focus within nano-nutraceuticals the programme will have, it better result in products that are technologically as advanced or more advanced than what Nestles and others will have on the market.

With the analytical techniques mentioned in this section completed, the competitive intelligence professional is now ready to make specific programme recommendations. Knowing the profile of targeted companies including risk orientation the analysts can make incentive recommendations. With the market analysis they can further refine the incentive. For example if the market is growing and profitable, a lower incentive rate should be made. If the target companies are highly risk averse and the opportunity is further out, a larger incentive would be recommended. In this hypothetical case with nano-nutraceutical research taking a long time and with a lot of regulatory uncertainty around it, as well as with concerns about consumer acceptance about nano-nutraceuticals a higher incentive will be required. Consistent with the market and profiles a recommendation of a grant or cash based incentive would be made over a tax credit as the grant or cash is more appealing when levels of risk are higher. Finally, the science mapping and timelining should provide the government with the intelligence needed to further target the incentive to those areas of nano-nutraceutical that provide better opportunities for Canadian companies.

11.4 Business Analytics

Business analytics is currently in vogue as a buzz word for the use of data to inform decision making in organizations (Davenport et al. 2010). In its Big Data incarnation, it is tied closely to the use of data mining techniques to analyze large complex data sets that might provide insights if mined properly. In reality, business analytics has been used in organizations for many years and hundreds of different techniques are available—all focused on optimizing one or more organizational outcomes. Ford Motor Corporation, for example, applied the basic notion of business analytics when, in 1914, Henry Ford decided to more than double employee wages. Conventional wisdom has it that increasing the cost of production will lead to increased prices and reduced demand. Ford, however, noted an increase in demand by approximately 60 % between 1914 and 1916 while prices dropped by 33 % during the same time frame. The sound application of business analytics enables managers to glean insights that might not be immediately obvious.

Analytic techniques might be categorized into three main types: "describe, prescribe and predict". Many organizations, in both private and public sectors are very good at descriptive analytics: charts and graphs about organizational phenomena such as how many companies took advantage of a government sponsored research credit programme where the companies were located, the amount of funds leveraged etc., Most however, are less capable of prescriptive analytics which could, for example, identify how best to allocate funds in order to optimize some organizational objective. Predictive analytics is being used in some organizations—one of the most mature areas being credit risk where, by analyzing a borrower's past behavior, income flows etc., accurate predictions about likelihood of default are possible. Predictive analytics has become the "holy grail", so to speak of analytics.

In policy development, the notion of "evidence-based policy" is founded on the idea of predicting the likely impact of policy interventions. At the moment, these predictions are subjective estimations. As will be discussed below, however, much is being done to better use data to make policy decisions.

11.4.1 Business Analytics in Government

Government organizations worldwide have embraced the notion of analytics. Evidence from the field of security such as passenger screening, tracking of aircraft, and the use of crime analytics to detect and ultimately prevent crime are all well-established applications of analytics in government organizations (Partnership for Public Service 2013). While the US appears to be ahead of many countries in applying analytics to the business of government, other countries such as Korea, Japan and Singapore have adopted risk assessment analytic approaches, intelligent traffic systems, and analytics driven monitoring systems to help anticipate and prevent occurrences such as epidemics and famine.

How does business analytics play into the scenario discussed above? If a policy initiative is to encourage businesses to invest in nano-neutriceuticals, a variety of analytic techniques can be used to anticipate the actual take-up of the provisions of the policy. We shall discuss two relatively simple techniques to illustrate the integration of foresight, competitive intelligence and business analytics within the context of national policies.

Two aspects deserve consideration: estimation of the expected value of the policy and evaluation of the likelihood that participants who are expected to avail themselves of the policy will behave in ways that will provide the expected value.

Econometric models are typically used to estimate the social and economic benefits to be derived from a policy. These models, however, rely on data gathered from stakeholders related to the policy environment. One relatively new approach to gathering data is "sentiment analysis". This approach, based on analysis of qualitative information appearing on millions of websites and blogs from the intended audience helps to identify opinions related to the outcomes being promoted by the policy. In addition to forecasting techniques such as scenario planning and roadmaps, sentiment analysis can provide guidance as to the attitudes prevalent in a particular population. It can be used for example, to predict the expected take up of provisions of the policy. Assuming that expected policy outcomes include the launching of businesses developing nano-neutriceutical products, analysis of consumer sentiment can provide clues of potential customer acceptance.

Two categories of analytics can be applied. Descriptive analytics would outline the percentage of posts that are positive or negative related to the policy in question. Based on this data, predictive analytics, using "Big-Data" techniques such as clustering for example, can separate the population who are posting about the policy into different groups based on characteristics such as age or geographic

Fig. 11.1 Clustering of
responses by age and
geography

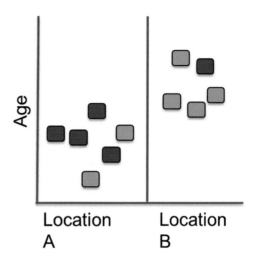

location. Based on this grouping, different simulations can be developed to anticipate the likely reaction based on changes to certain aspects of the policy. Figure 11.1
illustrates these ideas. Red icons are opposed to the policy, green icons support
it. The graph shows a clustering along age and geographic categories. In this case, it
indicates that aspects of the policy needs to be tweaked to better appeal to younger
people in location A. Depending on the information available about people in the
various cells (younger, older, location A, location B), data mining tools can be used
to predict whether such policy changes are likely to appeal to the each of the cells in
the graph.

Using such predictive concepts, government policy makers can anticipate how
the businesses they expect to participate in the initiative will respond. For example,
while intelligence through profiling might have identified 40 % as the likely rate for
the grant programme based on looking at past programmes and company profiles,
analytics will refine this by developing algorithms that will look at the risks
associated with the research and the companies risk attitudes.

With this information in hand, we can go one step further and simulate the
decision process used by *businesses* who might take advantage of the programme.
Businesses typically invest in new products in order to turn a profit. Investment
decisions can be quantified through the use of a variety of analytic models, one
being the "net present value" calculation. This approach discounts future expected
cash flows of an initial investment to estimate potential returns. The calculation is
as follows:

$$\$I - \left[\$R/(1+i)^t\right]$$

Where $\$I$ = initial investment, $\$R$ = expected returns (cash flow) from sales of the product, i = expected discount rate during the time period, and t = number of compounding period. As a practical example, supposed the business would need to invest \$400,000, for simplicity in this calculation let's say the government provided 50 % as part of the grant program they would get \$200,000 from the government, they would then have to get the remaining funds (assume it comes from internal funds) and they want to recoup that investment with an appropriate profit some point in the future. The company estimates cash flows for the first 3 years after full production of \$100,000, \$150,000 and \$2500,000.

Assuming a 6 % discount rate (added to the principal investment amount), the NPV works out to: \$469,931–437,743 (discounted cash flow) for a total of −\$32,188. With the government funds assume the discount rate is reduced to 3 % the \$400,000 investment is now \$432,358 leading to an NPV of +\$37,573. The decision rule for NPV calculations is to invest in projects with a positive cash flow (realizing that the decision maker might have a number of projects with positive cash flow).

Ultimately, using the formula to simulate a business person's decision model can help policy makers to better predict uptake and more accurately define the parameters of a policy.

11.4.2 Analytics Conclusion

A wealth of tools and techniques are available to policy analysts who would like to predict the likely outcomes of a policy. Data scientists, given the right type of data properly organized, can analyze and predict likely outcomes (Provost and Fawcett 2013). This capability will enable policy analysts to uniquely fine tune policies in order to improve the chances of realizing the results expected. As the technology available for Business Analytics improves and the techniques available to data scientists evolve, more government organizations will be making use of such tools to monitor and manage operational activities. Application of such empirically-based predictive approaches to policy initiatives is still in its infancy, but the opportunities are quite compelling.

11.5   The Combined Approach

Foresight and competitive intelligence contributes collection methodologies including primary and secondary collection approaches, facilitation methods, a variety of robust analytical methods, ability to work with qualitative information and a clear focus on understanding the external environment.

Business analytics contributes modeling capabilities, methods for dealing with massive amounts of quantitative data, emerging text analysis software for qualitative, a variety of proven internal indicators that have been used for dashboards and a rich history of primarily internal organization analysis with growing literature of customer analysis.

All three domains (foresight, intelligence and analytics) are used to provide decision making support and have complementary analytical techniques that allow the decision maker to better understand the external environment including key stakeholders. The combination of the three approaches is also useful for reducing the risk of designing a flawed program. Leaving any one approach out increases the risk of program failure as the program developer would be missing out on what could be a critical analytical component needed for program design.

The combining of the three approaches is starting to appear in some government programs. For example, in 2011, The United States government started a defence program called FUSE—Foresight and Understanding from Scientific Exposition. The program funds the development of "automated methods that aid in the systematic, continuous, and comprehensive assessment of technical emergence using information found in published scientific, technical, and patent literature." (Office of the Director of National Intelligence 2014). Specific research areas include text analytics, knowledge discovery, big data, social network analysis, natural language processing, forecasting, and machine learning. Small et al. (2014) reported on the program with a clear link between it and competitive technical intelligence. The techniques referenced and the automated analysis intent falls within the domain of business analytics, but the intent is to create foresight-based conclusions. This is a good example of the techniques from analytics and intelligence being needed for foresight purposes. In this example, the combining of the domains is crucial as each has analytical approaches that do not exist within the other but are necessary for getting this bigger picture.

11.5.1 Developing the Dashboard and Monitoring Mechanisms

The purpose of a program/policy dashboard is to give early warning of needed changes based on external environment signals. The initial nano-neutraceutical programme design was based on long term analysis through foresight, short and medium term analysis through intelligence and further refinement through analytics but in the end these techniques were used to reduce programme risks not eliminate them. Accordingly the idea of a program dashboard is to monitor the environment on an ongoing basis to ensure:

1. That we were correct on our analysis of the potential programme users and that the programme is being used as it was envisioned in the program designed by the appropriate users and for the appropriate technology development.
2. That we were correct in our analysis of the nano-nutraceutical market and that the underlying profitability, technology developments and so forth are consistent with what was assessed in the initial analysis. Also that the timeline that was projected for the industry also holds.

3. The longer term scenario and roadmap projected for the program design also holds.

A change in any of these could result in the programme not attaining its desired outcomes. Finding out about changes in any of these at an early enough point can lead to programme changes that will improve its effectiveness or lead to elimination of the programme if it is not meeting its objectives.

Developing the programme dashboard and its ensuing key performance indicators (what needs to be monitored) starts with competitive intelligence and timelining. In this case the program developer needs to look at all the activities (and measurements of activities) between announcing the programme and the final outcome, successful commercialized products and jobs. Based on our experiences with government programs these are the activities that are embedded in the timeline.

1. Program inquiries. Once a program is announced and before applications arise, it is normal for companies to ask local government officers for more information about the program, advice and counselling on applying for the program.
2. Applications/proposals: Eventually some months normally after the program is announced and after enquiries the companies submit the applications.

At this early stage, the first two steps as identified in an intelligence timeline exercise can be put on a dashboard and monitored (Table 11.4). For example, in Canada on a technical programme it would be expected that inquiries would be made to the National Research Council officer, to Canadian Business Service centres and in the case of some of the regions of Canada, regional development officers. As part of the dash boarding exercise, these individuals could be asked to enter into a database the emails received about the program and summaries of conversations. These could be analyzed using content analytics software. Applications/proposals when received can be subject to content assessment. Software could be used to look at the kind of development being proposed (is it what was envisioned by the program?). Analysis could also be done on the number of applications (is it at the desired level?) where the applications are coming from (regional distribution, type of companies etc). At this stage of the program, problems identified through analysis of dashboard data could be investigated and the program could be modified accordingly. Perhaps the program is not being advertised properly (we have seen this problem before) or perhaps the incentives provided are not significant enough to encourage the desired kinds of research. We have also seen situations where the incentive was appropriate but there was no interest in doing research in the area targeted by the program. This part of the dashboard focuses on validating and monitoring the company profiles.

Assuming the right companies are asking the right questions and applying for funds to develop the targeted technologies for commercialization, what comes next in the time line? What gets measured and dash boarded next? In this simplified example intermediary steps include:

Table 11.4 The nano-nutraceutical program dashboard

Early Outcomes

Advice and counselling

Inquiries

Applications

Intermediate Outcomes

R&D activity

University/college training programs

Business clusters forming

Patenting

Requests for commercialization funding

Long Term and Environmental Factors

Sentiment Indicators-Business

Sentiment Indicators-Consumers

Technology watch

Grey literature development

Timeline and roadmap milestones

Roadmap and scenario monitoring

1. R&D activity—Next should be hiring activities and R&D.
2. University/college training programs—If this is a new area of research appropriate labor availability should be an issue thus there should be a development of training programs to support the demands of the companies. Without the appropriate labor, the development and commercialization activities cannot occur.
3. Business clusters forming.
4. Patenting.
5. Requests for commercialization funding.

All the activities in the intermediate outcomes are needed for the final outcomes of the program to be realized.

The final item to place on the dashboard is monitoring of the external environment in terms of the underlying profitability, demand, interest, all the factors examined in earlier analysis. This is about monitoring the factors that underlie the strategic analysis, environmental analysis and evolutionary analysis. Environments do change and with it the rationale for the initial program. A few of the items that could be on the dashboard include:

- Sentiment indicators—business: this is an analytics approach assessing social media data for signs that companies interest in the targeted areas grows during the program.
- Sentiment indicators—consumers: this is an analytics assignment assessing social media for signs that consumers interest in the targeted areas (level 1 and 2 nanopharmaceuticals) grows during the program. Biofoods encountered a serious blow when consumer concerns dominated discussions.
- Technology watch, grey literature analysis, timeline and roadmap milestones: Governments need an ongoing technology watch program including grey literature analysis to examine if there are developments that were not expected. For example, perhaps another country has developed disruptive technology in the area or has done a high enough level of investment to move the technology forward at an accelerated rate. Developments need to be watched.
- Roadmap and scenario monitoring: similar to timeline and roadmap milestones there would need to be an ongoing effort to watch for signs of which scenario was emerging and milestones being met (or not met) on the roadmaps.

Information for the dashboard is generally readily available. The information required for the sentiment analysis could come from social media as well as assessment of emails sent to the government agencies about their programs. Grey literature analysis is well developed as an analytical discipline and appropriate databases would need to be accessed (these are all open source public databases). Technology watch, timeline, roadmap, scenario etc information would be gathered from several sources including:

- Ongoing foresight and intelligence projects. The government could on an ongoing basis do Delphi's to test the longer term assumptions and commission additional intelligence projects.
- Organized collection at conferences and trade shows. For example the Bio trade show would have workshops, booths, presentations and participants with the appropriate knowledge of developments in nano-nutraceuticals.

11.6   Conclusions

As mentioned at the beginning of this article, industrial policy is fraught with uncertainty due its reliance on external environmental elements for its success. Foresight, intelligence and business analytics taken together provide the toolkit to better understand this uncertainty and can help lead to more successful industrial policy. Foresight and intelligence with their external environment focus provide the tools to among other things understand the direction markets are going in, profile your local industry to determine what policy instruments can work on them, and better understand how technology is going to evolve. Signals picked up today through an externally focused intelligence effort can be used to confirm conclusions reached in longer term foresight initiatives such as scenarios, roadmaps and scans thereby providing the information needed to establish the longer term oriented industrial policy needed in science and technology related industries.

Acknowledgement The paper is a revised and developed version of a previously published paper by Calof J. et al. (2015) Foresight, Competitive Intelligence and Business Analytics—Tools for Making Industrial Programmes More Efficient. Foresight-Russia, vol. 9, no 1, pp. 68–81. DOI: 10.17323/1995-459X.2015.1.68.81 with permission from Foresight-Russia journal (http://fore sight-journal.hse.ru/en/).

References

Ashton WB, Klavans RA (1997) Keeping abreast of science and technology: technical intelligence for business. Battelle, Columbus

Bergeron P (2000) Government approaches to foster competitive intelligence practice in SMEs: a comparative study of eight governments. Proc Annu Meet Am Soc Inf Sci 37:301–308

Bonthous J (1995) Understanding intelligence across cultures. Compet Intell Rev 4(Summer/Fall):12–19

Calof J, Skinner B (1999) Government's role in competitive intelligence: what's happening in Canada? Compet Intell Mag 2(2):20–23

Calof J, Smith J (2010) The integrative domain of foresight and competitive intelligence and its impact on R&D management. R&D Manag 40(1):31–39

Calof J, Richards G, Smith J (2015) Foresight, competitive intelligence and business analytics—tools for making industrial programmes more efficient. Foresight-Russia 9(1):68–81

Davenport T, Harris J, Morrison R (2010) Analytics at work: smarter decisions, better results. Harvard Business School Press, Boston

Du Toit ASA (2013) Comparative study of competitive intelligence practices between two retail banks in Brazil and South Africa. J IntellStud Bus 2:30–39

Fehringer D, Hohhof B, Johnson T (2006) State of the art competitive intelligence. Competitive Intelligence Foundation, San Antonio

Fleisher CS, Bensoussan B (2002) Strategic and competitive analysis: methods and techniques for analyzing business competition. Prentice Hall, Upper Saddle River, NJ

For-Learn (2014) Excerpt from online foresight guide. http://forlearn.jrc.ec.europa.eu/guide/9_key-terms/foresight.htm. Accessed 8 Aug 2015

Fruchet G (2009) Effective practices for implementing CTI in large R&D organizations in competitive technical intelligence. In: Ashton R, Hohhof B (eds) The competitive technical intelligence book, vol 4. Competitive Intelligence Foundation, San Antonio, TX, pp 37–54

Gilad B (1998) Business blindspots. Irwin Professional Publishing, New York

Global Intelligence Alliance (2011) Market intelligence in global organization: survey findings in 2011. GIA White Paper

Global Intelligence Alliance (2013) The state of market intelligence in 2013. GIA White Paper

Juhari A, Stephens D (2006) Tracing the origins of competitive intelligence throughout history. Compet Intell Rev 3(4):61–82

Kahaner L (1997) Competitive intelligence: how to gather, analyze, and use information to move your business to the top. Simon and Schuster, New York

Office of the Director of National Intelligence (2014) Foresight and understanding from scientific exposition (FUSE). http://www.iarpa.gov/index.php/research-programs/fuse. Accessed 9 Aug 2015

Parker D (2000) Can government CI bolster regional competitiveness? Compet Intell Rev 11 (4):57–64

Popper R (2008) How are foresight methods selected? Foresight 10(6):62–89

Provost F, Fawcett T (2013) Data science for business: what you need to know about data mining and data-analytic thinking. O'Reilly Media, Sebastopol

Salvador M, Rodriguez R, Casanova LFS (2013) Applying competitive intelligence: the case of thermoplastics elastomers. J Intell Stud Bus 3:47–53

Saritas O, Smith J (2008) The big picture- trends, drivers, wild cards and weak signals. In: Third international Seville seminar on future oriented technology analysis, Seville, Spain, 16–17 Oct 2008

SCIP (2014) FAQ. www.scip.org/re_pdfs/1395928684_pdf_FrequentlyAskedQuestions.pdf. Accessed 9 Aug 2015

Small H, Boyack KW, Klavans R (2014) Identifying emerging topics in science and technology. Res Policy 43(8):1450–1467

Smith JE, Saritas O (2011) Science and technology foresight baker's dozen: a pocket primer of comparative and combined foresight methods. Foresight 13(2):79–96

Wolfe J (2005) Safer and guilt-free nano foods. Forbes Magazine 8/09/2005. www.forbes.com/2005/08/09/nanotechnology-kraft-hershey-cz_jw_0810soapbox_inl.html. Accessed 9 Aug 2015

Wright S, Calof JL (2006) The quest for competitive business and marketing intelligence: a country comparison of current practices. Eur J Mark 40(5/6):453–465

Xu K, Liao SS, Ki J, Song Y (2011) Mining comparative opinions from customer reviews for competitive intelligence. Decis Support Syst 50(4):743–754

Part III

Future Oriented STI Policy Context

Exploring the Potential for Foresight and Forward-Looking Activity in Horizon 2020

12

Jennifer Cassingena Harper

12.1 Introduction

Foresight, or forward-looking activity, has traditionally featured as a significant policy approach in the European Union's Framework Programme for Research, Technological Development and Innovation, ranging from its use to support accession countries and regions to thematic foresight. More recently foresight is being used to address grand societal challenges and smart specialisation. The approach has permeated the design, operation and content of the Programme, playing an increasingly prominent role in recent years in strategic policy and programme design, instruments and projects.

This paper focuses on the range of roles which foresight needs to undertake with the launch of the next Framework Programme, Horizon 2020. The context within which the Programme was designed and launched, within and outside the European Union, is disruptive and complex, requiring iterations of the programme to cater for the economic and financial crisis. The Programme now commands an even larger scale of resources and is focused on new more strategic approaches in terms of complex and long-term planning of different sub-programmes, prioritisation of societal challenges and focus areas, smart specialisation, as well as integrated approaches linking challenges and programmes.

The review of past foresight and forward-looking activity in the EU Framework Programme highlights the emergence of two broad types of roles, namely activities to support the take-up of foresight and forward-looking activity and strategic and instrumental use of foresight and forward-looking activity. A distinction is made between the strategic, instrumental and operational role of foresight in innovation

J. Cassingena Harper (✉)
Policy and Internationalisation Unit, Malta Council for Science and Technology, Villa Bighi, Bighi, Kalkara CSP 11, Malta
e-mail: jennifer.harper@gov.mt

© Springer International Publishing Switzerland 2016
L. Gokhberg et al. (eds.), *Deploying Foresight for Policy and Strategy Makers*,
Science, Technology and Innovation Studies, DOI 10.1007/978-3-319-25628-3_12

policy design and implementation and how this is reflected in Horizon 2020. The paper then reviews ongoing use of foresight in research and innovation policy at European level and in Member States which reflects a level of convergence which Horizon 2020 can capitalise on.

12.2   Foresight and Forward-Looking Activity at European Level

Over the last decade, foresight has gradually permeated policy design at regional, national and European level more and more extensively, supported by national and European programmes. The European Framework Programmes in particular from FP5 (Fifth Framework) to FP7 (Seventh Framework) have supported a range of foresight and forward-looking activities. These activities have extended from foresight to support accession countries develop their research and innovation policies, to regional foresight (sub-national level), international foresight, thematic foresight, and foresight to develop processes, tools and approaches tailored for European level rationales and context (Havas et al. 2010). These activities have taken on different forms including expert groups, support actions, networks, conferences and workshops, as well as studies and tenders, tailored to specific contexts, needs and constraints.

These activities can be clustered based on the rationales they were designed to address and the targeted outcomes and impacts (see Table 12.1 below).

It is interesting to note that significant investment was made in two broad types of activity:

Type 1: Activities to Support the Take-up of Foresight and Forward-Looking Activity
- Developing foresight guides which have provided an important reference for national and regional foresight. In particular the FOREN (Regional) Guide was identified as a high impact FP5 project, while the JRC Forlearn online guide provided easier access to information in designing and running a foresight exercise.
- Capacity-building and network development aimed at connecting different forward-looking communities in Europe (and worldwide) together.
- Mapping, scanning and analysis of forward-looking activity.
- Development of tools, processes and approaches tailored to European level.

Type 2: Strategic and Instrumental Use of Foresight and Forward-Looking Activity
- Address overall strategy and policy in the Framework Programme.
- Address thematic foresight.
- Support agenda-setting, vision-development, in different FP instruments including:

Table 12.1 Overview of foresight roles in Horizon 2020

	Horizon 2020	Role of foresight
Advisory/ Strategic level	Different core objectives, changing and refined over time: Exiting the Crisis, the Innovation Gap and the 3 % target	Rationalising different objectives, future-proofing through horizon scanning and alternative scenarios, strategic discussions through dedicated exercises
Instrumental level	Relating the focus areas to the challenges and not falling into trap of previous programmes	Priority-setting, exploring new approaches for bringing together different research and business communities, revisiting challenges and focus areas through embedded foresight
Operational level	Improving coordination and integration within EU and with and among Member States, particularly with outsourcing of certain Commission functions	Bringing together relevant parts of the Commission, creating synergies with Member State efforts in foresight

Source: author

- to support joint programming;
- to support joint technology initiatives;
- to support societal Grand Challenges approach;
- to drive international cooperation.

Type 1 activities predominated in FP5 complemented by thematic foresight for agenda-setting. In FP6 and FP7, Type 2 foresight has emerged more prominently, highlighting a growing awareness of the utility of foresight in a strategic and advisory role and in implementing a range of FP instruments which require different stakeholders to reflect on and agree on a shared vision and agenda for their joint activities. It is significant to note that stand-alone activity has gradually been complemented by more embedded activity linked to the policy cycle and addressing specific outcomes and impacts (see Table 12.1).

The foresight outcomes and impacts generated to date have been proven significant in certain cases, however greater coordination is required to improve impact. "(W)hile there are numerous forward looking activities both at European and Member State levels, these activities are uncoordinated and their results have a very limited impact on the actual preparation of policies and policy measures. We also need a shift from the thematic approach to challenge driven approach" (EFFLA Policy Brief 1).

Overview of rationales and impacts of European level FLAs

Type	Rationale	Examples	Targeted outcomes and impacts
Platforms and Networks	• Connecting communities • Access to Forward-looking information	Forsociety ERANET, EFMN/EFP, Forlearn, iKNOW, International Foresight Academy (IFA)	• Building networks, developing reference materials and fulfilling an important function as reference points
Studies	• Drawing "big pictures"	SCOPE 2015, Europe 2015, ESPAS, The World in 2025	• Significant impact and approach/results used in follow-up projects
Expert Groups	• Forward-looking advice on a broad range of possible topics	Regional Foresight (FOREN), Blueprints, Key Technologies, the World and Europe up to 2050, Europe in the world—AUGUR, Regional aspects and in particular on the Mediterranean area MEDPRO	• Considerable impact with results being used in related projects
Thematic I	• Focused foresight projects providing strategic intelligence "à la carte", but without embedding	Large number of projects, using a range of methods, e.g. Farhorizon, AgroFood, POLINARES, PASHMINA	• Significant impact with results used in design of programmes at European and national level
Thematic II	• Full-scale foresights embedded in strategy development • Clear need and demand to underpin strategy development using foresight	Very few examples: FISTERA, Freightvision, FOREN, eFORESEE and FORETECH	• FISTERA and FOREN considered high impact projects (FOREN in particular developed region-specific guides, FISTERA adoption of methodology at national level, contributions to EC policy)
Thematic III	• Strategy processes with a foresight component providing inputs to the definition of S(R)As	ESFRI, SCAR, JTIs, JPIs, ETPs	• High potential impacts (realised in SCAR foresight which led to significant follow-up activity)

Source: Adapted from presentation by Matthias Weber—extract from report prepared by the author for EFFLA in 2013 entitled EFFLA-commissioned Report on The Engagement of Member States (MS) in Forward Looking Activities (FLA) at EU-level

In the next section, the new Framework Programme, Horizon 2020, is analysed from the perspective of identifying how the role of foresight and forward-looking activity is likely to be shaped by the new context and the scale and ambition of the programme.

12.3   New Drive in Horizon 2020

The launch of Horizon 2020 marks a number of key changes in the European research and innovation ecosystem and its management. The context in which the changes are being undertaken is particularly significant, encompassing a range of shapers and drivers of European policy including:

- In terms of the backdrop, the highly dynamic, often disruptive, international policy context may not be a new factor in recent decades but the instantaneous impact of events as they happen in Europe and worldwide as a result of social media is wielding a disruptive effect on the shapers and drivers of policy at different levels and highlights the need for improved anticipatory and forward-looking capacity and more timely and yet policy responses which look beyond current crises.
- Societal concerns and demands and public opinion in the face of growing challenges linked to personal security, health, loss of confidence in societal structures and institutions, among others, is exerting a considerable pull on governance structures and services to reform quickly to new social realities and highlights the urgent need for tangible action to meet the current social reality through well-targeted investments in social innovation processes, products and services.
- The multitude of agendas and strategies which are being set at different levels creates a complex interconnected web of policy development and implementation with often overlapping if not conflicting goals and objectives. In the absence of effective joined-up governance processes, this calls for greater emphasis on managing the policy interfaces and the strategic levers for change and mechanisms for co-design of policy design and implementation.
- Upheaval and change in governance structures at all levels while required, to move towards more co-design frameworks allowing joined up policy-making, is not easy to manage effectively and requires a change in policy cultures, more openness and timely sharing of information and power.

In Horizon 2020, all these factors come into play. The financial and economic crises have exerted a strong effect in shaping the programme as it was being formulated, highlighting the need to anticipate and be prepared for sudden contextual changes. Horizon 2020 has as its core aims "responding to the economic crisis to invest in future jobs and growth, addressing people's concerns about their livelihoods, safety and environment and strengthening the EU's global position in research, innovation and technology".[1] Horizon 2020 is also to address more directly the Europe 2020 Agenda, the targets set therein (relating to R&D, as well as education, employment and climate and environment sustainability and fighting poverty and social exclusion) and the Innovation Union Flagship Initiative in

---

[1] http://ec.europa.eu/research/horizon2020/pdf/press/horizon2020-presentation.pdf.

Fig. 12.1 Interconnected flagship initiatives

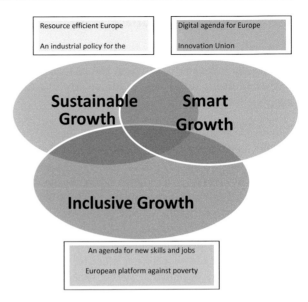

particular. It is important to note that the targets are "interrelated and mutually reinforcing: educational improvements help employability and reduce poverty; more R&D/innovation in the economy, combined with more efficient resources, makes us more competitive and creates jobs; investing in cleaner technologies combats climate change while creating new business/job opportunities."[2]

Similarly the Innovation Union Flagship Initiative is one of seven flagship initiatives, and the interfaces between these initiatives (see Fig. 12.1) require careful consideration and effective governance processes.

This is evident not least in efforts to bring greater coherence and joined up policy approaches across the related Commission Directorates-General (DG) in the area of research and innovation, and more broadly, the policy domains of regions, education, and information and communications technologies, enlargement and development among others. The outsourcing of project management and evaluations to external agencies to allow DG Research and Innovation to focus on strategic aspects of its mission has also been a key factor enabling the Directorate-General enhanced opportunities for embarking on a more strategically complex programme.

Indeed Horizon 2020 despite an outer packaging of increased simplicity and improved accessibility for applicants, reflects a new maturity and complexity in its design. The programme projects a new approach underpinned by strategic programming, based on the use of more evidence-based, coherent and future-proofed policy approaches. The programme brings together different financial instruments, the research funding programme, the Competitiveness and Innovation Programme

---

[2] Europe 2020 website:http://ec.europa.eu/europe2020/europe-2020-in-a-nutshell/targets/index_en.htm.

(CIP) and the European Institute for Innovation and Technology (EIT) as well as harnessing Structural Funds, Euratom and other European funds (education, COSME) and national research funding programmes. Managing this scale of activity entails serious efforts at joint agenda-setting, co-design of programmes, evaluation and combining a range of possibly competing (if not conflicting impacts).

12.3.1 Resolving Competing Rationales

The Horizon 2020 Programme thus combines different rationales set at different strategic levels and this calls for an ongoing dialogue between the managers of the different programmes and funding instruments as well as the range of beneficiaries and end users. This becomes more evident at the level of the sub-programmes which include:

- smart specialisation, targeting the regions;
- seven societal challenges, targeting the citizen;
- excellent science targeting the researcher and
- industrial leadership targeting companies.

The scale of the programme and the emphasis on multidisciplinarity, ideas to innovation, social innovation and integrated approaches again underline the need for more forward-looking, systemic and participatory approaches in designing and implementing the programme effectively. Indeed foresight features strongly in the design and implementation of the programme and there are efforts underway to secure a more prominent and reinforced role for foresight as the programme is underway (Fig. 12.2).

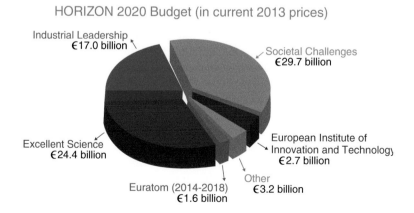

Fig. 12.2 Horizon 2020 budget. Source: Horizon 2020 Factsheet

This paper distinguishes between three levels, advisory/strategic, instrumental and operational, where foresight plays a role in supporting Horizon 2020. These different roles indicate the utility of foresight in shaping and defining research and innovation policy. In a recent NESTA report (Cassingena Harper 2013b), the advisory/strategic function of foresight has been related to efforts to rethink innovation policy and thereby restructure the innovation system as a whole or with a particular focus. The instrumental role refers to priority-setting, identification of opportunities, networking (e.g. industry-academia) and/or articulation of challenges and sub-challenges. In this paper, a third role for foresight is identified relating to its more operational function, allowing smooth transition processes towards co-implementation of policies and measures. These three roles are addressed in more detail in the next section, with a view to defining how foresight can be deployed more effectively through tailoring to particular levels and types of policy development and implementation.

12.3.2 Horizon 2020: Foresight in Strategic, Instrumental and Operational Functions

As indicated above, in Horizon 2020, foresight can and needs to be more prevalent and embedded in the different phases of policy cycle, from design, implementation, evaluation of impacts to iteration to move to the next round. Thus foresight has the potential and need to rise to the challenge by expanding its role to address three complementary functions, advisory/strategic, instrumental and operational (see Table 12.1 below). Apart from working within these policy levels, foresight needs to also play a role in creating interconnections between the three levels to ensure coherence and policy learning to inform successive rounds of work programmes up to 2020 and beyond.

Achieving the full potential of foresight will require investments in an enabling framework to address:

• Foresight training for staff tailored to specific functions.
• Awareness-raising on design and use of foresight in different contexts.
• Fine-tuning of foresight approaches and tools for particular context.
• Development of institutional mechanisms and facilities to support horizon scanning and sense-making.
• Drawing on good practices and expertise at national level and worldwide

Foresight has a key role to play at the strategic level in making sense of the range of objectives and goals to be addressed and the different programme rationales which need to be brought together and addressed as a whole, in consultation with all the stakeholder groups. This in turn creates the need to review current EU policy design structures and processes, and to assess if they are fit for purpose in this dynamic and complex policy context. The interdependencies and inter-linkages between the different agendas and programmes as well as possibly conflicting rationales require dialogue and consultation within a forward-looking prospective approach to allow a smooth transition to a desired state.

The strategic role of foresight in Horizon 2020 has been highlighted in the first Policy Brief (EFFLA Policy Brief 1) issued by the European Forum on Forward Looking Activities (EFFLA) which makes the following three key recommendations (Cassingena Harper 2013a):

- "Optimise the process of preparing the Commission's proposal for a future framework programme to be a process that draws systematically on inputs from Strategic Foresight. This needs to be done in the course of the next 3 years (i.e. in good time before the launch of the preparations of 'Horizon 2030').
- Set up a dedicated Strategic Foresight Unit or hub in DG Research and Innovation, in order to lead and coordinate the Strategic Foresight actions and embed them properly in the new strategy development process leading to the next proposal of the framework programme (to be regarded as a pilot for other strategy development processes).
- EFFLA recommends two enabling activities to be initiated in the short term. The first is preparing regular scans of the future starting now, in order to establish a rolling process of adjusting the annual work programmes of Horizon 2020 in a forward-looking manner. The second is the institution of a programme of Strategic Foresight training for EC staff."

This strategic level impacts on the national research and innovation agendas and programmes and on the instrumental level where foresight is used to support implementation of the programmes. This embedded function, though less prominent, can prove equally critical towards ensuring the effectiveness of the programme and in helping to secure desired outcomes and impacts.

In its instrumental role in Horizon 2020, foresight can prove an important lever in the implementation of the following policy instruments among others:

- Addressing the societal challenges and defining the sub-challenges.
- Smart specialisation in identifying the unique competitive advantage of regions and the related niches, in setting up public-private sector partnerships, triple helix networks and clusters.
- Demand side, e.g. public procurement for innovation supporting the client-provider relationships.

At the operational level, foresight can be used to support transition to new ways of operationalising policies and measures as required by the new context, in particular co-design and co-implementation. This includes:

- Governance processes aimed at improving policy harmonisation between EU and Member State level, including joint programming of national research programmes to address European societal challenges.

- Integrated approaches to address focus areas cutting across different challenges bring together all EU research and innovation funding (FP7, CIP and EIT) in a coherent, from-research-to-innovation overarching framework.
- New ways of operationalising demand side and business/society focus and social innovation.

In the next section, this operational function of foresight is explored in more detail as member states and the EU invest in joint efforts in Horizon 2020 to address grand societal challenges using joint programming and other initiatives. The paper highlights the need to identify ways of using member state foresight experiences to support European forward-looking activity in the new Framework Programme, Horizon 2020 and to invest in measures to support this type of cooperation.

12.3.3 Harmonising EU and Member State Forward Looking Activity

According to the NESTA Report on the Impact of Technology Foresight, Cassinga-Harper 2013b the review of Member States foresight activity linked to innovation policy highlights the fact that there has been:

> Mainly instrumental use of foresight, emphasis on priority-setting, networking and identification of opportunities. Impacts relate to informing the research and funding programmes and a focus on a defined set of challenges. At times the process and results have effects on innovation policy and strategy as a whole or lead to a higher level systemic foresight. Strategic/systemic foresight addressing the innovation system as a whole, can yield significant results depending on enhanced levels of preparedness, maturity and depth to the exercise and the prior identification and engagement of key persons and institutions who will directly use the results.

The review highlights particular strengths and features of Member State foresight with the German Futur Programme being particularly adept at bringing new actors into the strategic debate and introducing the concept of an 'informed public'. The UK Programme in its first incarnation succeeded in building industry-academia networks, while in its second round it helped to develop linkages across fields, sectors and markets or around problems. Scanning and exploring future opportunities to set priorities for investment in research and innovation (R&I) and identifying niche areas of competitive advantage has been the focus of foresight in several member states. The UK has been running a Horizon Scanning programme, while critical technologies exercises have been undertaken in France and Czech Republic.

Foresight on the national research and innovation system has entailed mapping, wiring up and enhancing the 'vitality' of the R&I ecosystem and the French FutuRIS exercise focused in particular on addressing the systemic challenges as well as identifying barriers to innovation. Foresight can have a more ambitious goal of strengthening the research and innovation ecosystem including building, transforming or reorienting the system, as in the case of the Hungarian and Swedish foresight programmes. In the case of Malta and Luxembourg, national foresight exercises led to the setting up of new R&I programmes and measures and the production of significant strategy/policy documents as was also the case in the third cycle of the UK Foresight programme.

In the accession countries, foresight is being used by the European Training Foundation (ETF) to support pre-accession countries to build on the sector approach in human resource development whilst operating in line with both their own human resource development strategy and the EU2020 strategy. The ETF FRAME project uses foresight to design more evidence-based and coherent national skills strategies, combining an emphasis on skills to be developed towards 2020, the role of the education and training system therein, the capacity needs of institutions to achieve this as well as indicators to monitor progress and mutual learning activities among the countries. "The countries seeking to accede to the European Union require a strategic approach to developing a vision for human resource development focusing in particular on the skills that are more likely to be needed in the period 2014–2020. In turn, a more coherent approach to pre-accession assistance also needs to be adopted, with result-oriented interventions tailored to the specific needs of each country and with clear targets and indicators to measure progress and achievements" (European Training Foundation FRAME Initiative). The initiative is distinctive in using foresight to address a range of local capacity-building needs including foresight and anticipatory skills, strategy and policy, institutional assessment and monitoring of human resource development, budgeting, targets, statistics, monitoring and impact assessment.

As smart specialisation has become a conditionality for accessing structural funds in the current programming period up to 2020, foresight is being used by a number of member states as a tool for bridging and harmonising national and regional (sub-national) research and innovation policies and strategies. Foresight has been used in Finland to design smart specialisation strategies, which complement national and regional innovation strategies based on identifying cross-cutting competencies and lead markets (Hermans 2015). The strategy is based on new innovation paradigms, focused on challenges and opportunities linked to value networks, building regional innovation platforms, developing a critical mass for a targeted number of multidisciplinary centres, many specialized centres and using national collaboration to share complementary competences across regions. The emphasis is on promoting related variety investing in cross cutting capabilities and foresight based demand driven business strategies among others and strengthening infant clusters to support long term growth (Havas and Keenan 2008).

In Lithuania, the national foresight process is being used in the design of the smart specialisation strategy with the main objective being "not to determine where to invest but how to help agents to discover where to invest in a decentralised and bottom-up logic. The methodology accepted in Lithuania departs from the traditional approach to priority setting focused on identification of research fields or economy sectors, and builds on the concepts of long term challenges and critical technologies. Choosing challenges-based priorities allows the better development of synergies and integrated policies, thus reducing fragmentation" (Paliokaitė et al. 2015). A mix of qualitative and quantitative methods approach is being used, including the expert panels, Delphi surveys, statistical and bibliometrical analysis, scenarios and roadmaps, and analytical studies.

In Malta, foresight approaches were linked to first regional innovation strategy exercise (MARIS) through the Futurreg project, an EU-supported Interreg 3c project, which in collaboration with MARIS explored the design of more demand-driven innovation policy with Maltese firms drawn from cross-section of foreign direct investment (FDI), traditional SMEs and start-ups. The results of this exercise which identified barriers to innovation related to lack of resources, knowledge and capabilities for innovation and the need for a culture for innovation (Cassingena Harper and Georghiou 2007), as an input for the current smart specialisation drive. Foresight has been embedded with this drive as reflected in the extensive, open participatory consultation processes with industry and other stakeholders to explore and confirm the sectors, followed by the setting up of sectoral focus groups. "In the discussions scenarios were explored within each of the regional branching pathways envisaged for smart specialisation whereby servicing of 'superyachts' represents a transition from an existing sector to a new one, the manufacturing sector is seeking to modernise, the tourism sector has a strategy to diversify through synergies with the cultural and education sectors and there are two cases of seeking to found a radically new domain" (Georghiou et al. 2014), namely digital gaming and exploitation of genetic and e-health data as a foundation for development as a venue for clinical trials and a biotech sector medicine capability.

This review of Member State use of foresight activity linked to national and regional innovation policy highlights a number of developments which are relevant for Horizon 2020 and the use of foresight therein. The more extended use of foresight to address competencies and capacities linked to national and regional economic priorities is significant, highlighting the growing importance of knowledge triangle approaches combining more coherently education, research and innovation strategies linked to specific economic niches. This is the territory which indeed Horizon 2020 is seeking to occupy and coordinate and by building on foresight at national and regional level, Horizon 2020 can ensure a sound base for its forward-looking activity, taking account of signals emerging from there for more grounded policies at local level. Smart specialisation can provide the key for improving the design of research and innovation policies at European level to meet better the needs and specificities of the range of member states and accession countries.

12.4 Conclusions

This paper has focused on the challenges facing Horizon 2020 as an ambitious, larger scale programme, combining varied rationales, objectives and targets. Foresight and forward-looking activity have played a significant role in the past in the design and iteration of European research and innovation policy and strategy. In Horizon 2020, what is required is a more targeted use of foresight at the strategic, instrumental and operational levels using dedicated exercises as well as embedded activity and complemented by capacity-building and support actions which develop missing competencies in anticipatory policy design and in facilitating the use of foresight approaches.

Building on the experiences of Member State and accession country foresight and EU level forward-looking activities, the paper highlights the need to develop ways of connecting these initiatives more closely and to develop ongoing platforms for sharing these experiences on a regular basis and having up-to-date information on ongoing initiatives. The improved coordination of forward looking activity at European level depends on Member States and accession countries being willing and able to share information, efforts and resources. By providing collaborative spaces, joint facilities and resources Horizon 2020 can ensure the tangible means for securing a coordination of efforts.

Finally foresight activity can only be sustained in the long-term if policy makers can be convinced of its utility in supporting policy and achieving desired impacts. The evaluation framework needs further work in order to improve the capture of impacts and this is an opportunity for Horizon 2020 to focus research efforts to develop a common evaluation and monitoring framework.

References

Cassingena Harper J (2013a) EFFLA-commissioned report on the engagement of member states (MS) in forward looking activities (FLA) at EU-level. http://goo.gl/nBDGkU. Accessed 6 August 2015

Cassingena Harper J (2013b) Impact of technology foresight. NESTA Working Paper 13/16. http://www.nesta.org.uk/publications/impact-technology-foresight. Accessed 6 August 2015

Cassingena Harper J, Georghiou L (2007) Using foresight to involve industry in innovation policy. EFP Brief No 187. http://goo.gl/fKpczo. Accessed 6 August 2015

EFFLA Policy Brief 1: Enhancing strategic decision-making in the EC with the help of Strategic Foresight. http://goo.gl/73mkXs. Accessed 6 August 2015

EFFLA Policy Brief 2: How to design a European foresight process that contributes to a European challenge driven R&I strategy process. https://goo.gl/VmDx0S. Accessed 6 August 2015

European Training Foundation FRAME Initiative. http://www.etf.europa.eu/web.nsf/pages/Frame_project. Accessed 6 August 2015

Georghiou L, Uyarra E, SalibaScerri R, Castillo N, Cassingena Harper J (2014) Adapting smart specialisation to a micro-economy—the case of Malta. Eur J Innov Manag 17(4):428–447

Havas A, Keenan M (2008) Foresight in CEE countries. In: Georghiou L (ed) The handbook of technology foresight. Edward Elgar, Cheltenham, pp 287–316

Havas A, Schartinger D, Weber M (2010) The impact of foresight on innovation policy-making: recent experiences and future perspectives. Res Eval 19(2):91–104

Hermans R (2015) Finland—smart specialisation based on cross-cutting competences and fore-sight lead markets. Presentation. www.observatorio.pt/download.php?id=612.Accessed 6 August 2015

Horizon (2020) Factsheet. http://goo.gl/AQlRlt. Accessed 6 August 2015

Paliokaitė A, Martinaitis Z, Reimeris R (2015) Foresight methods for smart specialisation strategy development in Lithuania. Technological Forecasting and Social Change, 101:185–199

Impact Analysis of Foresight for STI Policy Formulation: Cases of Romania, Vietnam and Kazakhstan

13

Ricardo Seidl da Fonseca

13.1   Approach and Method for Project Evaluation

The evaluation concept applied in the present paper is based on causal chain identification and analysis. The approach here pledges for a very clear and coherent differentiation between the actions and results controlled by the project itself, such as inputs, activities and outputs, and the actions and results eventually provoked or induced by the project but are fully dependent from the art and extension of the responses from the stakeholders or clients, such as outputs (or actual use of the outputs) and impact (as ultimate benefits) possibly attributed to the project (Fig. 13.1).

The second pillar of the subject evaluation approach is the application of the logical framework to identify the expected, projected or achieved results of the exercises, organized according to a hierarchy given by the related causal chain, and the circumstantial or structural conditions for their realization and evaluation (Table 13.1).

The defined logical framework would be used in the whole project cycle, from its design, through its implementation until its evaluation, which would influence possible adjustments in revising its design (Fig. 13.2).

The legal framework in this case should be used as an instrument for project monitoring and evaluation. It has to be seen as an on-work facility, being constantly completed and updated (Fig. 13.3).

Following this approach, the logical framework information shall be used to:

- guide and improve project implementation;

R.S. da Fonseca (✉)
Institute for Statistical Studies and Economics of Knowledge, National Research University Higher School of Economics, 20 Myasnitskaya Street, 101000 Moscow, Russia

Rua Pinheiro Guimarães, 115/1/1201, 22.281-080 Rio de Janeiro, RJ, Brazil
e-mail: r.seidl.fonseca@gmail.com

© Springer International Publishing Switzerland 2016
L. Gokhberg et al. (eds.), *Deploying Foresight for Policy and Strategy Makers*,
Science, Technology and Innovation Studies, DOI 10.1007/978-3-319-25628-3_13

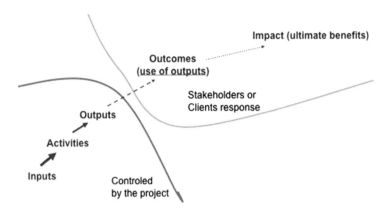

Fig. 13.1 Project causal chain. *Source*: author

Table 13.1 Basic logical framework

Result hierarchy	Performance indicators	Monitoring and evaluation	Assumptions and risks
Goal/Impact: Higher objective and impact to which a project, along with other processes, will contribute	Measure programme performance	The programme evaluation system	**(Goal to Strategic Goal)** Risk regarding strategic impact
Purpose/Outcome: Change in beneficiary behavior, systems or institutional performance because of the combined output strategy and key assumption	Measures that describe the accomplishment of purpose. The value, benefit and return of investment	People, events, processes, sources of data for organizing the project evaluation system	**(Purpose to Goal)** Risk regarding programme level impact
Output: The project intervention. The actual deliverables. What the project can be held accountable for producing	Measure the goods and services finally delivered by the project	People, events, processes, sources of data—supervision and monitoring system for validating the project design	**(Output to Purpose)** Risk regarding design effectiveness
Activities: The main activity clusters that must be undertaken in order to accomplish the outputs	**Input/Resources** Budget by activity, monetary, physical and human resources required to produce the Outputs	People, events, processes, sources of data and monitoring system for validating the project design	**(Activity to Output)** Risk regarding implementation and efficiency

Source: Thompson (2001)

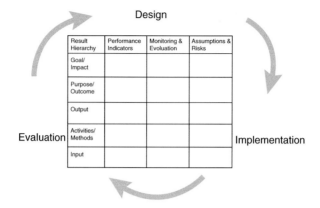

Fig. 13.2 Logical framework along the project cycle. *Source*: Thompson (2001)

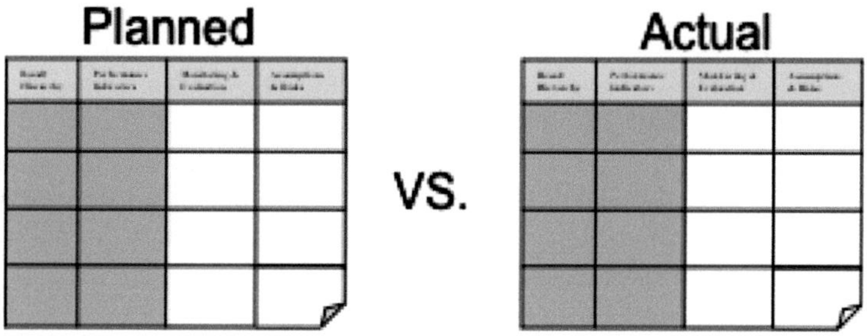

Fig. 13.3 Sequential use of logical framework. *Source*: Thompson (2001)

- assess overall project success at completion;
- improve future designs.

   Experience shows that a tide monitoring of a project with such instrument could lead to early identification and mitigation of most causes of default and unsuccessful results, as present in many post-factum project evaluation findings, such as for example:

- cost overruns or inadequate costing;
- need for extensions due to optimistic schedules;
- partially achieved objectives;
- lack of clear statement of objectives and performance indicators;
- lack of clear roles and responsibilities;
- unspecified or optimistic assumptions;
- no means for learning or adjusting to change;
- no clear statement of impact (social, environmental, technical, etc.), be short or long term.

In summary, both the application of logical framework and its use along the whole project cycle shall enhance the intrinsic quality of results and facilitate their evaluation. This is the main basis to the evaluation concept and praxis presented in this paper.

13.2   Challenges and Approach for Science and Technology Strategies: the Role and Impact of Foresight

Since the advent of the big science mode to organizing the systematic public and private support to science and technology (S&T) activities and related facilities, the science, technology and innovation (STI) systems are increasingly requested to present and demonstrate that their results are socially, economically and environmentally relevant (UNIDO 2010a, d). With the massive use of scientific knowledge in all those spheres, the responsibility of the STI systems has grown steadily.

The societal pressure on achievements from the STI systems is today and more in the future challenged by a series of developments and new situations, to mention some (Martin 2006; Seidl da Fonseca and Saritas 2005):

- new generic and converging technologies are dominating all new areas, thus implicating in a escalate higher impact on economy and society, moreover being more dependent on advances in basic research;
- shift from linear models to multidisciplinary and technologically complex dynamic ecosystems;
- move to open innovation;
- stronger and more sensitive relationship between science, technology, innovation and society;
- explicit longer-term policy for STI essential in the context of growing international competition.

Such challenges, among many others, are equally determinant for the future of the STI system and became almost impossible to meet through traditional S&T policy formulation (Havas et al. 2010; Gokhberg 2016; Kang et al. 2009). In this context, the weaknesses of this approach must be overcome, such as (Seidl da Fonseca and Saritas 2005):

- based on simple extrapolative prediction;
- narrow pool of expertise;
- passive outcomes: "white papers" or policy documents;
- limited ownership from the wider stakeholders;
- mostly normative proposals
- difficulties to predict disruptive and innovative solutions.

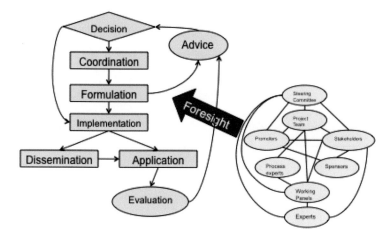

Fig. 13.4 Foresight as part of policy making. *Source*: author

13.2.1 Inclusion of Foresight in Policy-Making

Many authors and experiences suggest as an appropriate methodology for policy making the inclusion of foresight exercises as a structural component of the process of identifying, formulating and evaluating greenfield policies and strategies, addressing present and future challenges and issues (Fig. 13.4) (Destatte 2007; Georghiou 2003; Johnston 2012; Smith 2012).

Seidl da Fonseca and Saritas (2005) define such systemic approach as dynamic policy-making with foresight (Fig. 13.5).

13.2.2 What STI Foresight Should Achieve?

The basis for using foresight in defining future-looking policies for the STI system is given by the definitions offered by Ben Martin and Luke Georghiou:

- *Research foresight is "the process involved in systematically attempting to look into the longer-term future of science, technology, the economy and society with the aim of identifying the areas of strategic research and the emerging generic technologies likely to yield the greatest economic and social benefits" (Martin 1995).*
- *Technology foresight is "a systematic means of assessing those scientific and technological developments which could have a strong impact on industrial competitiveness, wealth creation and quality of life" (Georghiou 1996).*

These statements highlight the expected purposes, outcomes and goals as well as impacts of foresight exercises for STI.

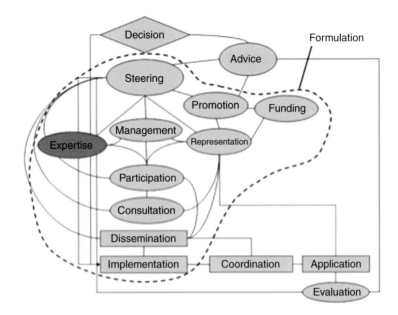

Fig. 13.5 Representing incorporation of foresight methodology as dynamic policy making.
Source: Seidl da Fonseca and Saritas (2005)

13.2.3 Input and Results of STI Foresight

Following the statements above, an analysis of a representative number of STI
related foresight exercises, permits to identify the categories of input and results
defined according to those of the logical framework. Table 13.2 exemplifies such
categorization. These categories are retained in the present modelling.

13.2.4 Focus of STI Foresight

Seidl da Fonseca and Saritas (2005) identify the major focus for STI foresight
projects as being policy, structural or S&T domains. Table 13.3 exemplifies the
particular focus according to these items.

13.2.5 Foresight Methods and Focus

Analysing actual experience from a representative number of foresight projects, as
well as the internal strengths of the different methods in use, Saritas and Seidl da
Fonseca (2005), propose a correlation between focus and most appropriate methods
for designing and evaluating those projects. Table 13.4 indicates such correlation.

Table 13.2 Foresight input and results

Input and results	Foresight
Input	• Foresight methodologies • Project focus, design and management
Output: tangible 'products'	• Critical technology lists • Baseline and benchmarking studies • Scenarios and future visions • Delphi survey databases • Anticipation and projections of long-term developments
Outcome: 'process' benefits associated with foresight	• Networking and resultant horizontal linkages • Commitment to guiding visions/recommendations • Adoption of long-term thinking and foresight practices >foresight culture
Impact: ultimate benefits—development goals	• Higher performance of the national innovation system • Emergence of new and competitive science-based products or services • Wider involvement of stakeholders in science and technology policies • Forging new social networks for shaping the future

Source: author

Table 13.3 Specifying focus

Major focus	Specification
Policy focus	• Priority setting • Identifying ways in which future science and technology could address future challenges for society and identifying potential opportunities and risks
Structural focus	• Reorienting or revitalizing Science, Technology and Innovation system • Bringing new actors into the strategic debate • Building new networks and linkages across fields, sectors and markets or around problems
S&T domains focus	• Identifying scientific trends, key technologies and clusters • Estimating impact and relations to industry, society and environment

Source: Seidl da Fonseca and Saritas (2005)

13.2.6 Criteria for Results Evaluation

Various foresight practitioners address the definition of criteria for conducting results evaluation of STI foresight projects. Table 13.5 proposes to allocate those criteria according to the categories of results as determined in the logical framework.

Table 13.4 Relating foresight methods with focus

Methods	Policy focus	Structural focus	S&T focus
Scanning	✓		✓
Bibliometrics			✓
Literature review			✓
Key indicators	✓		✓
Stakeholder mapping		✓	
System analysis		✓	
Megatrend analysis		✓	✓
Scenarios		✓	✓
Weak signals			✓
SWOT analysis	✓	✓	✓
Delphi survey	✓	✓	✓
Road mapping			✓
Relevance trees	✓		
Strategic planning		✓	
Critical/key technologies			✓
R&D planning			✓
Policy recommendations	✓		
Action planning	✓	✓	

Source: Saritas and Seidl da Fonseca (2005)

The present modelling develops the evaluation criteria according to the above mentioned categorization.

13.2.7 Impact Analysis Activities

On the basis of relevant experiences in realizing evaluation for foresight exercises in general, the present modelling proposes a sequence of research activities leading to a formalization and organization of the core and ancillary information composing the evaluation framework (Sokolova and Makarova 2013; Meissner 2012; UNIDO 2010b). The sequence of main research work[1] proposed by this modelling and used in the following evaluation of real projects, shall be:

- documentary analysis;
- stakeholders interviews;
- online stakeholders survey;

---

[1] For more details see Popper et al. (2010).

Table 13.5 Available categorization for criteria for results evaluation

Results		Johnston	CTFB	For-learn	Smith	AIT/Havas	PREST
Ouput	Formalization (report, book)		×				×
	Dissemination (workshops, newsletters, press articles, web sites)						×
	Analysis of trends and drivers		×				×
	Scenarios		×				×
	Strategic process		×				×
	Roadmaps		×				×
	List of key technologies and scientific areas		×				×
Outcome/ immediate impact	Develop connections and networks		×		×	×	×
	Integrate stakeholders into foresight programmes				×		
	Develop and employ methodologies and skills in wider circle				×	×	
	Sending opportunities and innovations				×		
	Create strategies				×		
	Awareness raising	×				×	
	Informing policy/ policy recommendations	×				×	

(continued)

- case studies;
- experts evaluation panel;
- benchmarking against similar exercises.

The following chapters exemplify the application of these impact analysis activities to evaluating three real STI foresight exercises.

Table 13.5 (continued)

Results		Johnston	CTFB	For-learn	Smith	AIT/Havas	PREST
Ultimate impact	Enactment of new STI policies and programmes		×	×	×	×	
	Creation of joint-ventures and STI agendas/new projects		×	×			
	Consolidation of research groups/institutions; consolidation of STI capacities		×	×			
	Emergence of social and technological innovations			×			
	Raising competitiveness through innovation					×	×
	Influencing wider policy, strategy, investment, programe delivery and public attitudes	×				×	
	Cultural changes towards longer-term and systematic thinking and addressing uncertainty	×				×	

Source: author

13.3 Evaluation of the Case Studies Romania, Vietnam and Kazakhstan

This section demonstrates the features of the modelling for STI foresight evaluation, where each of three real STI projects is firstly appraised to allow further a comparative evaluation. It starts with documentary analysis, supplemented with on-line interviews and survey. The final goal of the modelling is to constitute a framework for benchmarking of foresight work and methods applied in the formulation, implementation and evaluation of policy and strategies for STI development. The selected cases are few of foresight projects implicitly embeded in the process of STI policy making. The cases of Romania and Vietnam have been implemented in parallel and in interaction with the STI policy formulation. In the Kazakh case the foresight exercise follows the policy formulation, complementing it with the definition of the STI programmes necessary to realize the policy vectors. The presentation of the cases below proposes a template for STI foresight evaluation.

13.3.1 Romanian case: Romanian Public R&D System in 2020

The Romanian foresight was coupled with the preparation of the National Strategy for Research, Development and Innovation as well as the National Action Plan for Research, Development, and Innovation during the years 2007–2013. The foresight project aimed at supporting Romanian policy-makers' efforts to move towards more inclusive, collaborative, and future-oriented ways to develop public policies for the R&D sector, as well as to provide a negotiated equilibrium between distinct, and sometimes competing visions about the future. Overall the work took 1.5 years starting in June 2005 and being supervised by the National Authority for Scientific Research. Following the modelling concept, the Romania foresight had a clear policy focus, with the specification below:

- demand for scientific research and technology development;
- priorities for research, innovation and structural change:
  - information society technologies;
  - competitiveness through innovation;
  - quality of life;
  - social and cultural dynamics;
  - energy and environment;
  - structural change in the public R&D system;
- visionary objectives.

The foresight was developed as a parallel stream to the policy formulation, supervised by the same Steering Committee. Starting with an R&D analysis and backed up by negotiation workshops and conference, the information, findings and agreements generated under the foresight stream were incorporated into the policy formulation and further fed back the foresight. Figure 13.6 indicates how this methodology was conceived and applied. Figure 13.7 indicates the workflow and schedule of the Romanian project. As the main output of the whole exercise, in the first months of 2007, the National Action Plan for Research, Development and Innovation was adopted by the Romanian Government, and was passed as legislation. The R&D Programmes have been approved, incorporating:

- capacities;
- human resources;
- ideas (basic research);
- partnerships;
- innovation (exploratory, "out of the box" innovative thinking).

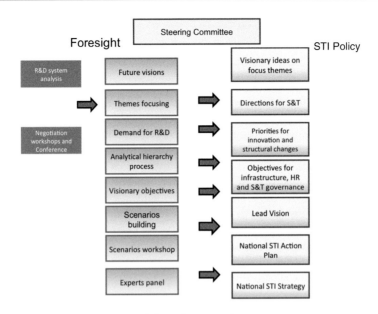

Fig. 13.6 Romanian foresight methodology. *Source*: author

Activities	06.2005	09.2005	12.2005	05.2006	07.2006	10.2006	03.2007
Work Package 1							
Management and Communication							
Work Package 2							
R&D System							
Work Package 3							
Foresight							
Work Package 4							
National Strategy for STI							
Work Package 5							
National Plan for STI							

Fig. 13.7 Romanian Gantt diagram. *Source*: author

According to Grosu and Curaj (2007) "the results of the foresight exercise were marked with high levels of legitimacy. They were widely regarded as a negotiated compromise about the future of public R&D sector. Both academia and policy makers accepted the results. They were considered not to describe "the best possible future", but rather "the most desirable future" for agents from most distinct systems within Romanian society.

The process brought to the forefront of a national dialogue on science and technology twenty-six fields of innovation. Results of research in four of these fields skyrocketed during the foresight process, and ever since, such as:

- Advanced information systems for e-services
- Advanced materials
- Quality of health
- Sustainable energy systems and technologies. energy security".

There is no evidence that the foresight process has functioned as a catalyst for creating networks of innovators.

Actually the foresight process pinpointed positive evolutions and phenomena, which were already on place. And this is its most important outcome.

13.3.1.1 Lessons Learned

Further from Grosu and Curaj (2007) the foresight process was focused on the part of the Public STI System that is involved in applied research. However, Romanian futurists learned very fast that one couldn't disrupt part of the system without disrupting the whole system. A parallel stream had to be organized, in which dialogue on the future of basic research was hosted. And even though there was little interaction between the two, this secondary stream also proved to be important. A good lesson to have in mind for future foresight exercises!

The outcome of the Romanian Technology Foresight solving this conflict is "undecided". Neither player of the original conflict was able to legitimize his solution through the foresight process, therefore conflict ended in a compromise.

13.3.1.2 Post Factum Evaluation

The foresight work showed an impact on the emerging R&D policy community in Romania and on the national STI community itself. There was a certain awareness creation momentum in both communities, which also led to policy makers moving along the learning curve towards understanding the STI community and the driving forces behind scientific progress and success. The Romanian Foresight was the first effort to identify long-term STI priorities of the country. The foresight assumed an increase in R&D expenditure initially but had to adjust the original plan due to the economic crisis in course of which spending was reduced significantly. Thus future foresight in Romania needs to aim at stronger long-term visions for the national STI system and measures to reach critical mass in priority fields. This will also require a more active involvement of all stakeholders.

Table 13.6 reproduces the opinions given by experts as replies to the on-line questionnaire distributed in connection to this modelling exercise. This activity corresponds to the phase of on-line interviews of the evaluation activities sequence.

13.3.2 Vietnam Case: Policy Advice to Science, Technology and Innovation Strategy (STI) 2011–2020 and the High Technology Law Implementation

The foresight exercise in Vietnam was dedicated to supporting the preparation of the STI strategy for the period 2011–2020 and the implementation of the High

Table 13.6 Summary of on-line interviews

Questions	Responses
Which other outcome and impact were/are expected to be achieved by the STI foresight exercise?	• An emerging R&D policy community (i.e. shared values, shared understanding of policy issues, common behaviour). • Increased awareness of the R&D communities about their position in the national and global R&D landscape. • Introduced futures thinking in policy-making processes. • An increasing learning curve associated with the policy formulation process (both at national and institutional level). In fact, the learning cycles that are enabled by the national STI foresight exercises highlight the different degrees of "maturity" of the STI community, which defines the innovation culture and which in turn is one of the crucial factors of success for all the items in the previous questions. To state it briefly, everything comes step by step.
In your opinion, which are the main failures of the STI foresight exercise?	• It did not offer a more diverse vision on the future of Romanian R&D. • The assumption that the public R&D expenditure will gradually increase from 0.3 % of GDP in 2005 to 1 % in 2010. The assumption was based on a governmental decision, but after the economic crisis Romania has been one of the few countries, which drastically reduced the public R&D expenditures. Hence the planning based on the foresight received for the period 2007–2013 only one third of the programmed funds.
What should be addressed or done better in new/further STI foresight exercises?	• To focus more on the nature of change. • To improve methodological rigor based on what was in fact implemented in the foresight exercise carried in 2013 for the National RDI Strategy 2014–2020. • A stronger stress on visioning at the level of the RDI system. • A broader stakeholders mapping and mobilisation using data-analytics technologies. • A thematic prioritisation based on the estimated necessary investment for reaching critical mass. • Better assessment of trends, cycles, opportunities, treats, and weak signals lurking in the future, as well as sense-making in a more cohesive fashion. • Better evidence based and data integration. • More diverse, inclusive stakeholders.

Source: author

Technology Law. The purpose of the foresight was to enhance the capacity of Vietnamese stakeholders in developing policies and strategies on S&T and industrial innovation compatible with the economic and social goals of the country (UNIDO 2010c). It took three years starting in November 2009. The work was supported by UNIDO and Austrian Institute of Technology (AIT) under the supervision of the National Institute for Science and Technology Policy and Strategic Studies (NISTPSS) and the Ministry of Science and Technology (MOST). Funding was provided the One UN Fund. The organization of the foresight involved a Project Management Unit (PMU), a team which was responsible to conduct the foresight exercise, a STI strategy team conducting the STI strategy 2011–2020 formulation process with inputs produced by the foresight team as well as international and national professionals specialized in respective areas of knowledge and participating in the workshops, and relevant stakeholders such as policy makers, industry, academy, university, research institutes think tanks, professional associations.

According to the modelling concept, the Vietnam exercise had a policy focus associated to a S&T domain from R&D to innovation. The description of the focus in this exercise shows the following structure:

- STI strategy to describe the approach for dealing with future challenges in the STI system:
  - Remediation of current shortcomings.
  - Anticipating future requirements.
  - A view on futures models for the STI system.
  - Define priorities.
- How to make R&D fruitful for innovation and thus for socio-economic development?
  - Not just R&D but also "innovation", i.e. how to make the step towards bringing new products and services to the market.

The exercise in Vietnam was planned to be conducted by two teams: one dedicated to the foresight component and another dedicated to the STI strategy, both to be developed in parallel. The same institution, NISTPSS, did the supervision. The foresight team would conduct the studies and provide the STI strategy team with defined results in papers. These papers shall constitute an immediate knowledge reservoir for the preparation of the envisage STI policy and strategies document. Figure 13.8 indicates how this methodology was conceived and applied.

In the case of the Vietnam exercise, the project document and the inception paper presented a predefined logical framework, which was used for monitoring the implementation process (Table 13.7).

13.3.2.1 Project Implementation

The exercise was conducted according to a two year basic schedule, being readapted during the real implementation. Figure 13.9 shows the projected work schedule.

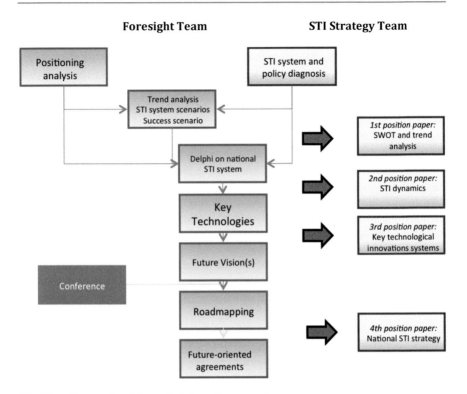

Fig. 13.8 Vietnam foresight methodology. *Source*: author

The exercise was intented to provide as output the reports and position papers, e.g. reports on results of methodologies applied, positions papers and an international conference connecting foresight process with the policy and strategy preparation. Furthermore the 5-year National Development Plan—NDP (2011–2015) and the MOST Science and Technology Strategy 2011–2020 (adopted in April 2012) can be accounted at least to some extend as output.

The foresight was used as a focusing and policy-informing tool which supported the preparation of a fully-fledged national STI strategy, facilitated the institutional embedding of the foresight and strategy process, established cross-membership between foresight and strategy development groups (institutional learning process), tested a new parallel process approach for foresight and policy design and consolidated STI strategy propositions in position papers by defining structural priorities for more efficient and effective STI system operation as well as thematic priorities, including key science domains, technology areas and application fields.

13.3.2.2 Lessons Learned

The social and economic developments that have taken place in Vietnam in the past years have provided a facilitating framework for a novel approach to STI decision-making, combining foresight tools with traditional programming methods. The

Table 13.7 Vietnam foresight logical framework

Results	Intervention	Indicators
Goal/ Impact	STI strategy to transform the Vietnamese economy to a sustainable and knowledge based one.	Government adopts a robust STI strategy 2011–2020, linking Vietnam's STI goals with its sustainable development goals.
Purpose/ Outcome	To enhance the capacity of Vietnamese stakeholders in developing policies and strategies on STI compatible with the economic and social goals of the country.	Number and coverage of participating institutions that indicate awareness and understanding of the foresight methodology as a powerful policy making tool.
Output	A draft STI strategy 2011–2020, formulated through a highly participatory and consultative foresight exercise on innovation and high technology.	1. Number and coverage of participating institutions contributing to the foresight exercise and STI strategy formulation. 2. Extent of stakeholder recommendations incorporated into the draft STI strategy.
Activities	• foresight methodology; • policy documents; • workshops; • conferences.	• number of professionals trained and involved in the exercise • conference proceedings; • foresight reports; • STI position papers.

Source: author

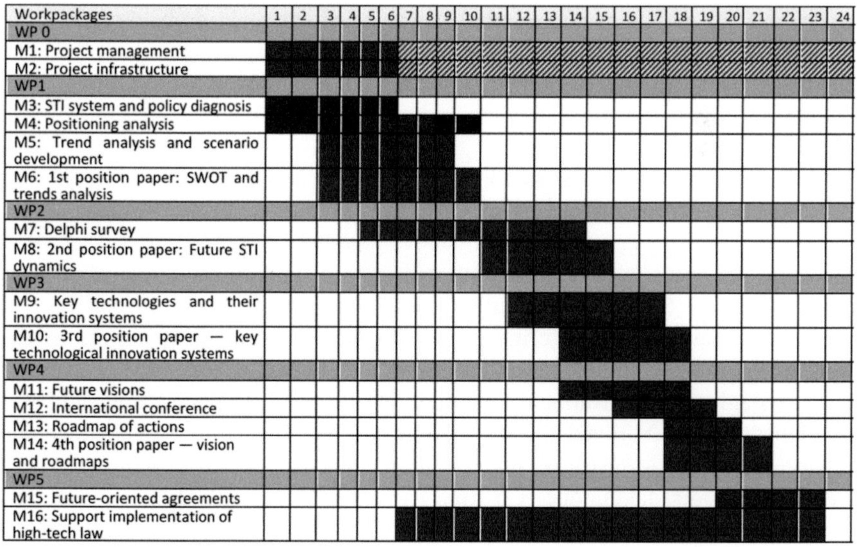

Fig. 13.9 Vietnam foresight Gantt diagram. *Source*: author

rather strong cultural context for policy definitions in Vietnam has limited the full application of the adopted methodological approach, but the process served as a powerful learning technique in the institutions dealing with policy and strategy. Because of the complexity in the definition of public policies in fostering and strengthening indigenous capabilities to use, adapt, modify or create technologies and scientific knowledge, a parallel process for foresight and policy design seems to be one of the most promising approaches to improve decision-making processes in developing countries (Aguirre-Bastos and Weber 2012).

13.3.2.3 Post Factum Evaluation

According to experts' opinion, foresight would expect to exert an influence on the formulation and concretization of the STI strategy 2020. Although some governance issues relevant to implementation of STI strategy were addressed, the actual implementation was not in the focus of the work. Instead the focus was on structural issues of upgrading the STI system, thematic (mainly technological) priorities as well as governance issues that could prevent (and have prevented in the past) the strategy from being effective. In all these regards, inputs were given, and also incorporated into the strategy. In this sense, the expected impacts could be regarded as partly achieved. However, the main problem consists of the difference between formulation and implementation. Foresight may have raised the point that there is also a need to ensure effective implementation and policy learning, but this does not seem to have found its way into the strategy and practice. Compared to initial planning, the foresight process could not be fully implemented due to numerous difficulties and problems encountered (some of which could probably have been anticipated). Table 13.8 reproduces the opinions given by experts as reply to the questionnaire distributed in connection to this modelling exercise.

13.3.3 System Analysis and Forecasting in the Field of Science and Technology in Kazakhstan

The foresight exercise aimed at the identification of the priority directions for the development of S&T in the medium-term, the formation of the portfolio of programmes and projects of critical technologies development within the framework of the priority directions. In this context, it should support the STI development strategies for the National Scientific and Research Councils. Foresight was developed by the National Centre of Science and Technology Evaluation and supervised by the Ministry of Education and Science, Ministry of Industry and New Technologies taking one year from December 2012 onwards.

The Kazakhstan foresight focused on STI priority directions specifying R&D areas on the basis of the main development vectors defined by the National Strategy 2020 in the fields of health, biotechnologies, new materials and technologies, safe, clean and efficient energy, environment and natural resources, information and communication environment, mechanical engineering, and sustainable development of agriculture, processing and safety of food.

Table 13.8 Summary of on-line interviews

Questions	Responses
Which other outcome and impact were/are expected to be achieved by the STI foresight exercise?	• Stakeholder's thinking about future may be impacted during STI-foresight process, routine in the way they think about the future may be challenged and changed; more solutions for present problems would come from the future's point of view; public could have a broader vision for seeking a standing flow of the events throughout certainties and uncertainties. • The main outcome has been the understanding of the importance of a foresight process in planning. • Learning effect among scientists and experts. • Awareness of importance of science and technology.
In your opinion, which are the main failures of the STI foresight exercise?	• How to combine between the fragility of foresight exercise's outcomes and the vital mission of decisions? In the most cases, how to persuade policy makers to apply foresight's recommendations in practice is always very difficult and challenged. • The difficulty is the lack of experience and interest of participants in the exercise to respond to questionnaires (e.g. Delphi). • Delphi bottom up approach is not in Vietnamese mind set. • The exercise was too ambitious for the limited competencies and experiences available in the country, both at the level of the implementing organisation (NISTPASS) and the stakeholders that were supposed to be involved in the process. • The embedding of the exercise in MOST was useful, but too limited for having a major impact. Other ministries than MOST were not involved to the extent necessary in order to ensure an impact on their STI-related agendas, which are very significant elements of the national STI System. The same holds for those actors at regional and sector level that are supposed to implement the strategy. • Participation was not effective for a number of reasons, ranging from lack of familiarity of participants with the methods to cultural reluctance to respond to questionnaires. • Myopic view from a part of stakeholders. Hard to get longer-term perspectives.

(continued)

Table 13.8 (continued)

Questions	Responses
What should be addressed or done better in new/further STI foresight exercises?	• More emphasis on training, less demanding methods and approaches to start with. • Broader involvement of other policy stakeholders (ministries, regions, etc.). • Much better preparation of the stakeholder engagement (media campaigns, user panels, PR & communication, etc.). • More attention to those governance issues that prevent foresight from being effective. • Adapting foresight methods to what can realistically be expected from the contributors. • Regular exercise is necessary. • More methods can be employed. • International exchange of the foresight results. • Future exercises should involve more expert groups, or focus groups techniques. • It would be more practical, if we could choose a framework and techniques that would be more suitable for a context of developing countries, like Vietnamese one. • To be more focused at sector level. • To involve the private sector.

Source: author

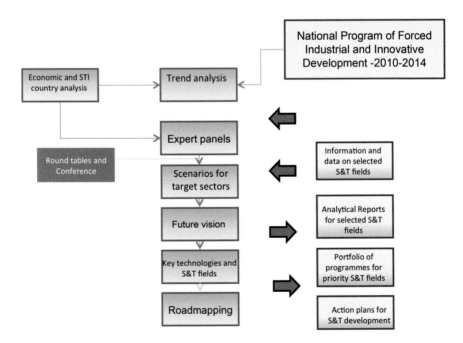

Fig. 13.10 Kazakhstan foresight methodology. *Source*: author

The foresight was timely placed between the publication of the main policy document, the National Programme of Forced Industrial and Innovative Development 2010–2014 and the preparation of the Action Plans for S&T development addressing the priority directions. The methodology established an exchange of documents and information both on the terms and conditions of STI development in the country and the foresight findings. Figure 13.10 shows this methodological concept.

It was expected that the foresight results in expert groups' reports on technological forecasting according to the priority STI directions in the medium-term, a list of the key scientific issues and one on critical technologies as well as proposals for a portfolio of related programmes and projects, and roadmaps for development of the latter. Furthermore the foresight should have contributed to the development and implementation of the state policy in the field of STI, industry, environment, health and safety, strategies or programmes with action plans for the necessary changes to the existing legislative framework, formation of infrastructural, personnel and financial support.

The foresight should ultimately contribute for the achievement of the goals stated at the policy document "Strategy 2020" that Kazakhstan becomes one of the 50 most competitive economies in the world by 2020 with a diversified economy and increased integration into the regional and global economy, with participation in international innovation processes.

13.3.3.1 Post Factum Evaluation

The foresight was the first of this type in Kazakhstan and proved impactful on awareness creation for long-term STI priority setting. It also showed that there is an urgent need to place STI-related matters higher on the agenda of national policy makers and to incorporate related priorities in the national mid- to long-term financial planning. Also priorities as defined in the national strategies need to be broadened with stronger emphasis on social sciences and related fields. Table 13.9 summarizes the opinions given by experts as reply to the questionnaire distributed in connection to this modelling exercise.

13.4   Comparative Analysis

The application of the proposed modelling for foresight exercises evaluation to the three case studies allows conducting a comparative analysis. Firstly, according to the available documentary analysis, reference is made to different foresight methods used in each of the cases, as indicated in Table 13.10.

To note is that among the various methods applied (Romania with 10 methods, Vietnam with 13 and Kazakhstan with 9), although the exercises are similar, only three methods are common to the three exercises: system analysis, scenarios and STI policy recommendations.

Table 13.9 Summary of on-line interviews

Questions	Responses
Which other outcome and impact were/are expected to be achieved by the STI foresight exercise?	• Guides direction of the whole R&D efforts of the country. • This was the first national S&T foresight public impact of showing that priorities should be made in order to increase the impact of S&T on national economic and social success.
In your opinion, which are the main failures of the STI foresight exercise?	• Results are often not taken into account in the planning and decision-making. • Ownership of the foresight should include people from highest floors of S&T governance. • Experts are asked to join with their expertise. Usually, experts are to think that their R&D items are very important. Thus, they are inclined to indigenously and originally stick to their ideas, which can lead the whole results distorted. Thus, it is quite important how to design STI foresight frame that leads those participating experts join the foresight professionally in their expertise and objectively at the same time.
What should be addressed or done better in new/further STI foresight exercises?	• More diversity of expertise including humanities, societies, economies, etc., should be involved. • This exercise was almost totally an expert project, even if a very large number (200 experts) participated. Next time a broader participation should be considered.

Source: author

For future STI foresight design, a better coordination between the focus of the exercise (policy, structural or S&T) and the most appropriate methods for designing and evaluating those exercises, as proposed in Table 2.3 should be considered.

Secondly, still according to the documentary analysis and using the template proposed in this paper, Table 13.11 below indicates the results achieved or to be achieved by each of the cases.

Examining Table 13.11, allows to draw conclusions

• Whereas the Vietnam case indicates a full range of outputs to be achieved, the other cases have a poor outputs' coverage; the only common output is "Lists of key technologies and scientific areas".
• The Kazakhstan case does not indicate any outcome to be achieved and both other cases indicate different and few outcomes.

Table 13.10 Foresight methods applied in the three projects

Methods	Romania	Vietnam	Kazakhstan
Scanning/focusing/positioning	×	×	
Bibliometrics/literature review			×
Experts panels	×		×
Key STI indicators	×	×	
Stakeholder mapping	×	×	
System analysis	×	×	×
Megatrend analysis		×	
Multiple criteria analysis			
analytical hierarchy process	×		
Scenarios	×	×	×
Future visions/lead vision	×	×	×
Weak signals			
SWOT/STEEP analysis		×	
Delphi survey		×	
Road mapping		×	×
Relevance trees			
Strategic planning			
Critical/key technologies		×	×
Future-oriented agreements		×	
STI policy recommendations	×	×	×
Action R&D planning/programming	×		×

Source: author

- In terms of achieved or expected impacts, all three cases assume only one or two items, which contradicts the main purpose of the exercises towards policy and strategy formulation and implementation.

Such results shall be an alert to STI foresight designers to enhance the project concept, methods and management, in view of constituting a robust instrument for policy making.

The modelling proposed in this paper uses an on-line questionnaire to be filled out by the different stakeholders for checking the main achievements, realized or expected, of the STI foresight exercise. The questionnaire constitutes of three blocks.

The first one shall allow analysing how the results of the foresight exercise contributed or would contribute to enhancing STI policy and strategy making. The related items are the following:

- producing and publicising prospective or long term studies;
- providing inputs or elements to priority setting and planning;

Table 13.11 Balance of results from the three cases

Results		Romania/ 2007	Vietnam/ 2012	Kazakhstan/ 2014
Output	Formalization (report, book)		✓	✓
	Dissemination (workshops, newsletters, press articles, web sites)	✓	✓	
	Analysis of trends and drivers; scenarios	✓	✓	
	Strategic process and roadmaps		✓	
	Lists of key technologies and scientific areas	✓	✓	✓
Outcome	Develop connections and networks		✓	
	Integrate stakeholders into foresight programmes		✓	
	Develop and employ methodologies and skills in wider circle			
	Sending opportunity, innovation; create strategic direction	✓		
	Awareness raising; managing conflicts and opposed views	✓		
	Informing policy/policy recomendations	✓	✓	
Impact	Enactment of new STI policies and programmes	✓	✓	
	Creation of joint-ventures and STI agendas/new projects; consolidation of research groups/institutions; consolidation of STI capacities			
	Emergence of social and technological innovations			
	Raising competitiveness through innovation			✓
	Influencing wider policy, strategy, investment, program delivery and public attitudes			
	Cultural changes towards longer-term and systematic thinking and addressing uncertainty; foresight culture	✓		

Source: author

- grade of influence and power of the mobilized stakeholders to actually shape the future through endogenous decision-making;
- fostering the development of competitive innovation-oriented industries.
- enhancing governance of the innovation system;
- adapting/up-grading public R&D institutes to the requirements of future innovation system.

The second block shall analyse the grade of engagement of the most relevant stakeholders and knowledge sources (experts and scientists). The categories proposed are:

- government (executive) authorities;
- legislative authorities;
- funding institutions;
- public S&T institutions;
- higher education institutions;
- private STI laboratories;
- business sector (public or private);
- labour force representations;
- independent scientists or experts.

The third block shall verify the targeted impact of the foresight exercise in the dimensional and performance indicators related to the STI sector. The selected indicators are:

- gross domestic expenditures on R&D (GERD);
- GERD as a % of GDP (GERD/GDP ratio);
- total R&D personnel;
- business enterprise expenditure on R&D (BERD);
- number of patent applications;
- number of innovating enterprises;
- R&D-intensive exports;
- high education expenditure on R&D (HERD);
- number of doctoral graduates;
- number of publications indexed by ISI databases;
- number of technology centres, R&D facilities and test labs;
- number of .com and .org domains;
- digital access index;
- public knowledge about S&T through exposure to mass media.

All items are to be assessed by the involved stakeholders (beneficiaries, policy makers, sponsors and experts) according to the grade given on which extent the foresight exercise combined with the preparation of STI policy and strategies, contributes to achieving the related result. The realized or expected achievement shall be estimated with a grade 0–5 (0 = no result; 5 = the highest result).

Table 13.12 Mean and standard deviation of grades given to achievements

Item nr°	Item	Romania		Vietnam		Kazakhstan	
		Grade	σ	Grade	σ	Grade	σ
How the results of the foresight exercise contributed to enhancing policy and strategy making?							
1	Producing and publicising prospective or long term studies	3,33	1,37	3,8	0,75	5	0
2	Providing inputs or elements to priority setting and planning	4,83	0,37	3,8	0,75	4,5	0,5
3	Grade of influence and power of the mobilized stakeholders to actually shape the future through endogenous decision-making	4,28	0,69	3,2	1,32	4,5	0,5
4	Fostering the development of competitive innovation-oriented industries	3,66	1,1	3,4	0,8	3,5	0,5
5	Enhancing governance of the innovation system	4,4	0,8	3,8	0,74	4	0
6	Adapting/upgrading public R&D institutes to the requirements of future innovation system	3,83	0,68	3,8	1,47	3,5	0,5
Achieving a broad participative process, engaging most relevant stakeholders and knowledge sources from:							
7.1	Government (executive) authorities	4,28	0,69	3,2	0,4	4	1
7.2	Legislative authorities	3,14	0,63	2,2	0,75	2,5	0,5
7.3	Funding institutions	4,14	0,83	2,6	1,02	5	0
7.4	Public S&T institutions	4,28	0,88	3,6	1,02	4,5	0,5
7.5	Higher education institutions	4,28	0,45	3,2	0,98	4	1
7.6	Private STI laboratories	3,42	1,05	1,75	0,83	3	2
7.7	Business sector (public or private)	3,14	0,63	2,6	0,8	4,5	0,5
7.8	Labour force representations	3,14	0,99	0,8	0,75	2	1
7.9	Independent scientists or experts	4,71	0,45	3,2	0,75	4	0
How indicators of STI and social-economic development were/would be affected by the STI-foresight exercise?							
8.1	Gross domestic expenditures on R&D (GERD)	4,28	1,03	2,8	1,47	3,5	1,5
8.2	GERD as a % of GDP (GERD/GDP ratio)	4,14	1,35	2,8	1,47	3,5	0,5
8.3	Total R&D personnel	3,42	1,59	2,4	1,02	3,5	0,5
8.4	Business enterprise expenditure on R&D (BERD)	3,28	1,03	2,8	0,75	3,5	0,5
8.5	Number of patent applications	3,57	1,05	2,4	0,8	2	1
8.6	Number of innovating enterprises	3,42	0,72	2,4	0,8	4	0
8.7	R&D-intensive exports	3,28	0,7	2,6	1,02	3,5	0,5
8.9	High education expenditure on R&D (HERD)	3,57	1,34	2,4	1,02	3,5	1,5
8.10	Number of doctoral graduates	3,71	1,48	2,2	0,75	3	1
8.11	Number of publications indexed by ISI databases	4,57	0,49	2,4	1,02	2,5	0,5
8.12	Number of technology centres, R&D facilities and test labs	4,14	1,35	2,6	1,2	3,5	0,5
8.13	Number of .com and .org domains	3,16	1,95	2,2	0,75	2,5	0,5
8.14	Digital access index	3,28	1,16	2,5	1,12	2,5	0,5
8.15	Public knowledge about S&T through exposure to mass media	4	1,07	2,6	1,01	5	0

Source: author

With the available questionnaire on-line a number of stakeholders provided their assessment on the realized/expected achievements for the three cases, which results are indicated in Table 13.12. Attributing a value lower than grade 3 as non-relevant achievement and equal or higher than grade 4 as an important achievement, such values are indicated in the table respectively in red and green. The modelling assumes that standard deviation of lower than equal to 1.0 could indicate consensus between the respondent and divergence if it is higher than 1.0, and in this shall be considered carefully. The proposal here is to apply those ranges of values to benchmarking the cases under evaluation. Using the three cases to exemplify the meaningfulness of the questionnaire's results, the following statements and findings could be proposed:

1. In all three cases stakeholders opinions expresses relevant to important contributions of foresight to enhancing policy and strategy making, where for Romania and similarly for Kazakhstan such role carries more strengths than for Vietnam.
2. The Romania exercise shows an ability to mobilize almost all relevant stakeholders, with a higher presence of the traditional actors of the S&T sectors and less, but still relevant, participation of the actors of the innovation sector.
3. In the Vietnam case, only the traditional stakeholders of the S&T sector could be mobilized at a relevant level, with exception of funding institutions, and the R&D innovation sector presents a low or non-relevant participation.
4. In Kazakhstan, almost all stakeholders show high participation, although legislative and labour sectors' participation is non-relevant.
5. The achieved or expected sensitization of performance indicators through STI foresight is very disperse as noticed among the three cases, where no common pattern can be seen.
6. In Vietnam, it seems that the foresight exercise has low relevance to affecting the performance of the STI sector, although the exercise was supposed to be conducted in close cooperation with the STI planning process, showing a low adherence between the two exercises.
7. In Romania, on the contrary the figures show a high adherence between foresight and policy making as almost all indicators are importantly affected.
8. The Kazakhstan case suggests, that only the number of innovation companies and public awareness on S&T could be influenced by the STI exercise, with low or non-relevant expectations in affecting all other indicators.

The few examples above demonstrate already the usefulness of the presented questionnaire as a component of this modelling proposal.

13.5   Conclusions

The studies and empirical research made to prepare the present modelling proposal could identify and contribute to a meaningful formal impact analysis of STI foresight. However, many fundamental questions arise on the effective embedding of foresight into the decision making process. In this sense, in responding the online questionnaire, It is pointed out:

> As a general remark: it is very difficult, if not to say impossible, to make any sensible statements about higher-order impacts of the foresight, such as the ones asked for in the (questionnaire) set of questions. Assuming that the STI strategy, which was at least partly underpinned by the foresight, is going to be implemented as planned (which is rather unlikely), the impact on most of the indicators would be—hopefully—mildly positive. However, there are so many conditions and other intervening factors involved that any substantive answer is highly problematic, even more so when we refer to long-term impacts (UNIDO 2010c).

Thus analysing long-term, high edge impact of foresight exercises, especially those embedded to STI policy and strategies, will make real meaning in so far as the formulation and implementation of the exercise realizes the whole cycle from idealizing to realizing.

References

Aguirre-Bastos C, Weber M (2012) Foresight and STI strategy development in an emerging economy: the case of Vietnam. EFP Brief No. 246

Destatte P (2007) Evaluation of foresight: how to take long term impacts into consideration? FOR-LEARN mutual learning workshop evaluation of foresight. Brussels, IPTS-DG RTD, 19 September

Georghiou L (2003) Evaluating foresight and lessons for its future impact. The second international conference on technology foresight, Tokyo, 27–28 February

Gokhberg L (ed) (2016) Russia 2030: science and technology foresight. Ministry of Education and Science of the Russian Federation, National Research University Higher School of Economics, Moscow

Grosu D, Curaj A (2007) Technology foresight in a power-oriented strategy for change. Conclusions after the Romanian technology foresight exercise. Futura 2:52–66

Havas A, Schartinger D, Weber M (2010) The impact of foresight on innovation policy-making: recent experiences and future perspectives. Res Eval 19(2):91–104

Johnston R (2012) Developing the capacity to assess the impact of foresight. Foresight 14 (1):56–68

Kang MH, Lee LC, Li SS (2009) Developing the evaluation framework of technology foresight program: lesson learned from European countries. Atlanta conference on science and innovation policy. http://hdl.handle.net/1853/32392. Accessed 7 Aug 2015

Martin B (2006) Strategic decision-making in policy formulation and the role of foresight. UNIDO training programme on technology foresight, module 3: technology foresight for decision-makers. Budapest, Hungary, 9 November

Meissner D (2012) Results and impact of national foresight-studies. Futures 44(10):905–913

Popper R, Georghiou L, Keenan M, Miles I (2010) Evaluating foresight: fully-fledged evaluation of the Colombian technology foresight programme (CTFP). Universidad del Valle, Santiago de Cali

Seidl da Fonseca R, Saritas O (2005) Instruments for strategy and policy: modelling the structure of the policy-making on science and technology. 14th International conference on management of technology, Vienna, Austria, 22–25 May

Smith J (2012) Measuring foresight impact. EFP Brief No. 249

Sokolova A, Makarova E (2013) Integrated framework for evaluation of national foresight studies. In: Meissner D, Gokhberg L, Sokolov A (eds) Science, technology and innovation policy for the future—potentials and limits of foresight studies. Heidelberg, pp 11–30

Thompson RM (2001) Logical framework training guide. Team Technologies Inc., Springfield, VA

UNIDO (2010a) Inception report. Doc. STI-WP0-MOD2-001-v7-010610

UNIDO (2010b) The science, technology and innovation system and policy analysis. Doc. STI-wp1-MOD3-001-V.4-020610

UNIDO (2010c) Trend analysis and scenario development of the Vietnamese STI system. Doc. STI-WP1-MOD5-012-V.1

UNIDO (2010d) Technology Foresight Manual, UNIDO, Vienna

The Future of Services

14

Ian Miles

14.1   Introduction

"Service" is a central concept in many fields of economic, management and policy thinking, as well as in professional and technical areas (e.g. "service marketing", "service management", "service-oriented architecture"). Contemporary societies are widely described as service economies—which lends a little more substance to the description than simply calling them "post-industrial societies". The service *sectors* comprise the bulk of employment and value-added in most OECD countries. Just over 70 % of employed people in the EU-27 worked in the service sectors in 2012. But also, service *occupations* have risen to the fore across all sectors of the economy, with huge shares of the workforce engaged in white-collar, front office, sales and distribution, and other activities. Over 40 % of employed people in the EU-27 in 2012, for example, were skilled non-manual workers (e.g. legislators, senior officials, managers, professionals, and associate professionals, technicians), while less than 25 % were skilled manual workers (e.g. plant and machine operators, craft workers) and less than 10 % in "elementary occupations".[1] The service industries and occupations span a huge range, just as do service industries. Some service industries (e.g. health, education, design, professional services) involve intellectually demanding work and have many highly qualified employees in professional occupations; others (e.g. hotels and restaurants, retail services) involve a great deal of routine work and frequently feature

---

[1] Data from Teichgraber (2013).

I. Miles (✉)
Manchester Institute of Innovation Research, University of Manchester, Oxford Rd, Manchester M13 9PL, UK

Institute for Statistical Studies and Economics of Knowledge, National Research University Higher School of Economics, 20 Myasnitskaya street, 101000 Moscow, Russia
e-mail: ian.miles.manchester@gmail.com

© Springer International Publishing Switzerland 2016
L. Gokhberg et al. (eds.), *Deploying Foresight for Policy and Strategy Makers*,
Science, Technology and Innovation Studies, DOI 10.1007/978-3-319-25628-3_14

low-skilled workers in often precarious occupations. Given this range of activities, discussing the future of services seems to involve a huge agenda.

But this is just the beginning. For example, many firms in manufacturing and elsewhere take their "product services" (advice, after sales and services complementary to the material product—training, software related to hardware, etc.) very seriously. Numerous manufacturers see such service *activities* and *products* as having displaced the focus on their traditional material product, as providing a competitive edge. This has given rise to an ugly neologism—servicisation (or servitization), the increasing offering of services as outputs.[2] Note that these services may not always be product services—often we see firms offering to sell "process services", their capabilities in R&D and design, in running information networks, in managing logistics, to others. So services are not just produced by service industries—these just happen to be the industries which are seen to have services as their main products. The term "services" may be used to describe service industries, or the products of these industries (and of servicised firms in all sectors).[3] But the meaning of "services" can also encompass various levels of **granularity**. "Services" can refer to very specific transformation *activities* in business or public organisations, or the whole package of outputs that are provided to the customer/user.[4] (For example, "hotel services" might refer to a host of activities from changing bed linen on—or to the entire hotel experience—or to the firms providing these experiences.)

To add more scope for confusion, there is a revival of the use of "service" to refer to benefits that have been provided to people. "Eco-systems services" are those valuable things provided by natural systems—such as breathable air! Even when limiting ourselves to economic activities or the products of human labour, the perspective that "service" is the ultimate objective of these activities has reoriented a great deal of management philosophy. This transformation has been summed up as the creation of a new "service dominant logic" (Vargo and Lusch 2006). This logic challenges many established approaches and practices, and has been a point of view that has captivated many active in service marketing and related areas. However, managers and engineers concerned with the design and operation of systems for producing or processing, say, biomaterials, chemicals, metals, and microsystems, tend to be less enamoured of its contribution. The take-away message that economic activities are ultimately producing services is nevertheless a valuable one, and resonates with the point made by Gershuny and others some decades ago—that

---

[2] For a study using survey data to explore the trends, see Dachs et al. (2012).

[3] So services produce services—which gives more room for confusion than saying, for example, that manufacturers produce goods.

[4] In this essay we will, with some reluctance, use the term "customer" most frequently. The term is often contentious—professional services may involve "clients", health services "patients", transport services "passengers", and so on—and it can be misleading in that the recipient of the service need not, furthermore be the person or organisation that is the paying customer for it; and some services (e.g. prison services) are changing the circumstances of certain individuals in order to obtain social outcomes for others.

people may access similar service outcomes in different ways. They may purchase some services from private providers, or have access to public services, or generate their own "self-services" using goods (equipment and consumables) that they have acquired. For example, I might treat a headache by seeing a private doctor, calling a public service health helpline, or taking an analgesic—or I could opt for alternative approaches such as having a neck massage or practising mindfulness meditation. Gershuny's insights led him to consider that the future of services was intimately bound up with innovation processes and with people's choices as to how to use their time and money. The relative price and quality of alternative modes of service provision, and of the goods (and complementary services) required for self-servicing, are liable to shape patterns of expenditure and of both formal employment and informal work in the "service economy".[5] Note also that some of the human labour involved in producing services can be "informal"—not just the consumer's effort that may go into coproducing a service with a service organisation, but also the effort that may go into producing "self-services" through using domestic equipment like cars and washing machines.

If the concepts of service and services are complex ones, so will thinking about the future of services be. This essay will attempt to avoid complicating matters further, and will eschew any attempt to forecast the likely trends in service employment, productivity, professions, and the like. It will draw on ideas developed in the course of studies of service innovation, and use the sorts of approaches applied in foresight studies to discuss how these ideas can help us think about the future of services in a meaningful way. It is no substitute, of course, for a comprehensive foresight study of the future of services (or of particular service activities)—that would require engagement with numerous experts and stakeholders who could bring together their understandings of how the various features of services and the wider terrain they exist within are being and could be reshaped.

14.2   Borders of the Service Vista: Specificities, Commonalities, and Convergence

Bitner (in 1992; and in subsequent work with various colleagues) introduced the resonant term "servicescape", to describe the physical setting in which service delivery takes place. The design of, for example, cinemas, hospitals, lecture theatres, restaurants, can have a large impact on the customers' service experience. Given that "servicescape" is already in use, we here will use the neologism "service vista" to signify the wider landscape of service activities.

This section of the essay addresses how far services share common features. Are there some core elements of services that make it really meaningful to discuss, for example, the services sector of the economy as if it were composed of activities with common and distinctive properties? The next section will address some of the

---

[5] Gershuny (1978), Gershuny and Miles (1983); for work on time use, Gershuny (2000).

main contours of the service vista—different types of service activity, in particular, and the changes that are underway in them.

Traditionally, services were not of great interest to economists—what they had in common was the view that they were of little importance. Despite the signifi-cance of, for example, transport and associated insurance in enabling the growth of world trade from the seventeenth century on, service activities were generally treated as an unproductive drain on the **real economy**—a view that continues to surface with some frequency.[6] In part this may have been because they tend not to produce physical goods that can be stored and amassed into visible repositories of wealth; in part because many services could be dismissed as worthy but extra-economic activities (religious services, housework), or as consumer luxuries like entertainment. Early efforts to produce systems of national accounts typically treated service industries as the residual that is left over when we have listed the productive sectors of agriculture and extractive industries, manufacturing and construction, power and water utilities, and so on. But as service industries became more important, efforts began to be made to identify common features that could be used to characterise them in more positive ways.[7] Many of these features relate to the key characteristics of intangibility and interactivity.

Intangibility is often seen as a common characteristic of service products. The products are rarely material goods as such. While primary industries are mainly extracting materials from the natural world; and secondary industries are mainly *making things* (goods and buildings) from these raw materials; tertiary industries—service sectors—are mainly *doing things*. "Doing things" means effecting transformations. These may be transformations in the state of artefacts (things like goods and buildings), or in the state of people or data. Note that some transformations actually involve preserving things or people, preventing other forces from transforming them (maintenance services, storage, preventative medi-cine, etc.)

Service products' intangibility leads to problematic features—for conventional analysis, at least. One of these is the difficulty in storing and transporting these products. Often physical presence is required: the service supplier has to move to be close to the user, or vice versa. This feature, sometimes known as coterminality, is closely related to the blurring between the acts of production and consumption. This has led some commentators to say that the service process is the product—for example, a massage or theatrical performance. (This forgets the training and rehearsal that may be involved, however.) The blurring of production and con-sumption, together with the difficulty of inspecting the product in advance is a factor in the "information asymmetries" that may exist between service supplier

---

[6] There may be some justification in the growing juxtaposition of the "real economy" and those financial services that seem to be dealing with layer upon layer of abstract financial products.

[7] See Miles (1993) for an extensive list of features that characterise service activities and products to greater or lesser extents. In that paper the point is made that just about any effort to generalise about characteristics of services will meet with many exceptions. This could be labelled Miles' First Law of Services.

and customer. The quality of the service may be hard to for the consumer to judge before it has been delivered—which is a rationale for many service activities to be regulated in one way or another, since the assumptions of transparent information in a free market are being violated. Another related feature is the likelihood of difficulties in assessing the productivity of service organisations and their employees; though we will not discuss this issue at length, there is a good case for thinking that there are often deep flaws in the statistics used to support claims about low productivity in many service activities.

There are many service activities that do involve more tangible elements. I recently spent an hour in a dental surgery, during which time a small block of ceramic was cut by a milling machine to the precise shape that the dentist had generated using Computer-Aided Design software and 3-D imaging of my gum and the stub of the old tooth; the ceramic crown was glued onto the stub of the old tooth and I walked out with a tangible, if artificial, new tooth. Similarly, repair and maintenance services may rebuild devices, paint or rustproof oil platforms. Information-related services may result in physical reports or computer software being produced, and hard cash handed over for these. Restaurants serve solid meals and liquid drinks. Trade services deliver purchased goods to consumers—they do not produce these goods,[8] but they do store, display, and provide access to them. Even though service activities are more about doing things than about making things, the things they do can make tangible things happen.

Earlier, we noted the shift from purchased services to self-services produced with the aid of goods. While much of the economic growth in the postwar period was related to consumer appliances—cars, washing machines, etc.—based on motor power, there are also cases where the shift is associated with information technologies (electronics and more recently microelectronics). An obvious example is the shift in terms of expenditure and time use towards more consumer activity with TV sets and music systems as compared to cinema and theatre shows. New computer-communication systems allow for the intangible information components of many services to be delivered online, so that physical presence becomes less important. Thus medical advice can be obtained at home, financial transactions undertaken via ecommerce, and money itself withdrawn from cash machines rather than the traditional bank branch. Some of the aspects of intangibility that have limited trade in services may thus be rendered less intractable.

The intangibility of many services is associated with another key feature, which we can label **Interactivity**. This captures the point that service activities typically involve more of a relationship between supplier and customer than is the case for other activities (where production and consumption are typically quite separate). Terms such as "consumer-intensity" (Gartner and Reissman 1974) and "servuction" (Eiglier and Langeard 1987) are deployed to signify the interaction between service supplier and user. Interactivity may take different forms, and be more or less intense

---

[8] Though goods may be assembled from components and/or customised, for example in bicycle shops, and tailors.

at different stages of the "service relationship" or "service journey" (which may involve the customer and service organisation coming into contact at a number of "touchpoints", and even in a variety of different servicescapes). For some services, interaction is most intensive at early stages in the process (where the nature of the customer requirement is being established) and later stages (when the service is delivered), while the actual service production activity tends to be more of a back-office affair. Thus a design firm or advertising agency may do a lot of the production work on its premises, before providing the customer with the proposed solution to their problem (Doroshenko et al. 2014). For some services, in contrast, interactivity is much more intense when the production, delivery and consumption are largely coterminous; the presence (and appropriate behaviour) of the consumer is needed for personal transport or entertainment to be provided. With the exception of a few craft and professional services, though, there will often be a back-office operation ensuring that the customer can be provided with the appropriate service: in the case of transport there are major logistics operations involved in making the vehicles and their staff available and operational at the right time; for entertainment services there may well have been extensive rehearsals, stage design, and the like. In some cases the customer is mainly required to provide their physical presence; in other cases there may be a good deal of information exchange needed for the service to fit the customer. The term "self-service" was used before Gershuny seized on it, too, to describe how some service organisations involve some customers in the work of selecting among choices—whether in a cafeteria or at a cash machine. Another related idea is "coproduction" (e.g. Bettencourt et al, 2002), which sees both supplier and user as producers of the service product, which would not be created without active engagement on both sides. This will be particularly significant when the user is expected to be more than a passive consumer of the service—when they are expected to behave in particular ways, such as following a medical regime, participating in a sports activity.

The (more or less) intangible service that is provided as a result of (more or less) interactivity is often described, too, as being relatively highly tailored or customised to the specific customer. While service suppliers regard quite a large share of their output as being standardised, this actually varies a good deal across different service industries.[9] Some manufacturing firms do specialise in small batch production, and construction projects may be very one-off, but in general there seems to be a strong case that most service activities involve relatively more customisation than do most secondary sector activities. Some service industries, especially the more knowledge-intensive ones like health services and KIBS, are particularly likely to produce services that are more specialised to particular

---

[9] Evangelista and Savona (1998) and Hipp et al. (2000) found many service firms reporting standardised outputs, in their surveys; conflicting results are, however, reported by Sundbo (2002) and Hortelano and González-Moreno (2007). These differences may reflect variations across countries, time periods, or survey methods.

customer needs. These are also activities that often require much interactivity and coproduction of the service.

The implications and dynamics of interactivity and intangibility are often related, for instance we could attribute coterminality to the frequent need for service supplier and client to be at the same place at the same time (though, as noted, the use of information technology may reduce this for some service activities). Likewise, productivity issues arise again. Interactivity implies that service quality will be a matter not only of supplier effort, but also of the clients' own inputs. This means not only that the value of the output may depend on the consumer, but also that productivity as measured by labour inputs by the supplier may be achieved at the cost of more labour required from the user. Productivity assessment also becomes difficult because of the **heterogeneity** among the outputs of a service organisation, with many services being bespoke or at least are customised. The heterogeneity of outputs contributes to the difficulty in assessing service quality prior to service production, and to the difficulties confronting service productivity measurement.

In discussing both intangibility and interactivity, we can see changes over time associated with technological developments and with new strategies and business models on the part of service providers—and sometimes, in all likelihood, with changing consumer demands (e.g. away from packaged holidays to personalised experiences). In thinking about the future of services, the issue of how far service organisations seek to exploit and built on these two characteristics, and the extent to which they try to overcome any limitations they may imply, is one of the big questions. For example, will the organisations pursue economies of scale with more standardised services and use of new IT to internationalise their operations; or will they seek to "move up the value chain" and focus on more specialised and costly services (which, if internationalised, would demand quite different sorts of international communications)? Or, rather, which subsets will pursue which combinations of these strategies?

In pioneering work on service industries, Levitt (1976) drew attention to the industrialization of services. He was an early exponent of how service firms were expanding by establishing more branches, producing essentially the same service in more places to more customers. They were standardizing (elements of) their products. This was accomplished through assembling the final service products in an economically efficient manner, often with a high division of labour and the use of a lower skilled workforce (consider, for example, the creating of meals from standardised tangible elements in the case of fast foods.) The creation and combination of the component modules would often involve mass production techniques and more capital intensity (i.e. reliance on technology). Recent discussions of "McJobs" reflect the tendency of firms following this model of service industrialisation to rely upon relatively high levels of low-wage and fairly unskilled staff, often working on a part-time or insecure basis.[10]

---

[10] For a rare treatment of the beneficial possibilities of such arrangements—with citations to more negative accounts—see Gould (2010).

More than a decade after Levitt identified the trajectory of service industria-lisation, Miles (1987) argued that we should view this development in the light of the growing awareness that industrialism was itself entering a new phase (variously described as "new times", "post-Fordism", the "third industrial revolution", and so on). A major feature of this new phase was seen the shift of much manufacturing industry away from traditional mass production. In response to pressures exerted by trends in consumer demand, and opportunities offered by the application of flexible IT and new organisational approaches, mass production models were being replaced by "mass customisation". Customisation used to imply that the product would be created on a one-off basis for one specific customer[11]; but "mass customisation" could involve a production line flexibly adjusting to create numer-ous variants on the same basic product to suit specific customer specifications. Each car on the line, for example, might have different colours, seat fabrics, entertain-ment systems, and so on. Without going into the details of this model, two points were noted in the context of services. First was that manufacturing might be coming to resemble service activities more, for example in heterogeneity and customisation of products, in interaction with customers (at least about product characteristics). Manufacturing firms were also producing more service products alongside their goods (though the term "servicisation" was not yet in use). Second was that service industrialisation was not just a matter of service activities being organised like traditional manufacturing mass production. It also involved forms of "mass customisation", so that standardised products were liable to come in many varieties with rapid change from customer to customer. Miles (1987) thus wrote of the "convergent economy", where many of the features supposedly demarcating services and manufacturing were becoming less definitive.

In the terminology employed in the present essay, the borders of the service vista have become more porous; and the future of services may resemble the future of manufacturing in many respects. Indeed, the grand sectors are so intertwined, as we shall see, that they are inseparable parts of a shared future, though understanding just how they may be brought together requires attention to the details and specificities of services.

The characteristics common to many services, discussed briefly above, meant that many services were organised on a craft basis, in small firms (or, in contrast, sometimes in large public service organisations). These small service firms are still prevalent in many industries, and a typical structure of a service sector is for a large number of small firms (usually serving very local markets) to coexist alongside a relatively small number of large and often transnational bodies. (Some large firms organise themselves through a franchise model, so that local "branches" are in effect independent business, tightly constrained to follow corporate marketing, quality control and other standards. IT is also quite common for smaller service firms to band together in some kind of association that allows them to share

---

[11] Though there might be many units produced for that customer—e.g. many mobile phones produced with specific logos on them and software in them, for a particular network operator.

common costs.) It is still the case that most service industries feature a larger share of small firms than do manufacturing industries (think of a small shop versus a small factory). A few service activities really require massive networks and scale economies (railways, airlines, banks), and some of these are organised or highly regulated by state authorities (health, education, criminal justice). Larger service firms have, however, emerged in industries as diverse as hotels and restaurants, on the one hand, and advertising and computer services, on the other. The same corporate presence is apparent in many cities around the world, in consequence.

These large firms have found ways of overcoming some features of service products and production processes that made it harder to achieve economies of scale, to reach out to international markets, and to escape the constraints of bureaucratic organisations and conservative professions. They have been aided by socio-political changes that have liberalised many economies, opening up professions and public services to competition, including services into trade agreements and rules. And they have stimulated and been facilitated by organisational strategies and application of new technologies and related innovations.

Again, we can view many of these innovations as a matter of overcoming the characteristic challenges of services. Does intangibility make the service difficult to communicate or recall? Add tangible elements—like loyalty cards, concert memorabilia, and certificates. Is it difficult to demonstrate, or to assess the quality of the service? Establish free trials or entry-level versions of the service (as in freemium services); provide evidence of quality through certification and self-regulation user group support, etc. Does interactivity makes the role of the customer central? Make use of user communities as ways of supplying service content or otherwise enhancing the service; encourage users to behave in ways that enhance the experience of others; develop forms of self-service that save supplier costs while enabling more consumer choice. Need to intensely exchange information? Use new Information Technology to allow for anytime, anywhere communication and access to automated elements of the service; to support the quality of learning and enable reproducibility of delivery through better presentation systems. Innovations of these sorts are pervasive in modern services, and are rapidly evolving—not least as new and more powerful devices supporting interactivity and information exchange are becoming available. Service industries have become much more technology-intensive, as Richard Barras anticipated[12]; and they have been closely related to new technologies owned by (or at least surrounding, or carried by) consumers, as our lives have been restructured to take account of new technological opportunities. Becoming more technology-intensive is another way in which service industries resemble manufacturing more closely (and much of the work at least looks similar, with computer screens and keyboards near-universal). But also, the innovations discussed here are ones that are often emulated by manufacturing firms.

---

[12] Barras (1990) and earlier studies on the "reverse product cycle" where he argued that IT should be seen as the basis for an industrial revolution in services).

Whether you are interacting online with the sales department of a manufacturing firm or with a wholesale or retail firm may not be immediately obvious, and service and nonservice firms' websites and social media presence are strikingly similar. Again, this can be viewed in terms of convergence, not just of one sector coming to sound more like the other, but of both sectors acquiring a new common vocabulary. But it also draws our attention to the intertwining of goods and services, with service products being at least partly produced and delivered by means of advanced technological artefacts—those themselves rely on both hardware and software. We can anticipate that the future of services will be intimately bound up with future technological innovations—not because technology "impacts" on services or social life, but because new knowledge of how to effect transformations in things, in people, in data, offers opportunities for social actors to improve their circumstances. The results that ensue depend on how different actors, with their different material and cognitive resources, are willing and able to act on this knowledge.

14.3   Contours of the Service Vista: Varieties of Services

Already, reference has been made to various distinctive service industries—for example mention has been made of KIBS, which fall into two broader categories— knowledge-intensive services (along with health, education, telecommunications, etc., which do not only serve businesses) and business services (along with office cleaning, secretarial services, building security, etc., which are not particularly knowledge-intensive). KIBS themselves can be disaggregated into T-KIBS (those based on technological knowledge, such as computer and engineering services), P-KIBS (professional knowledge such as accountancy and legal services) and arguably C-KIBS (creative or cultural knowledge, such as advertising and design services).

A good place to provide a broad view of the contours of the servicevista is via the statistical classification of services. While service industries were treated in very broad-brush terms in early national accounts, this classification has been given much finer detailin recent decades. The current NACE (revision 2) system (Eurostat 2009) runs from "section" G—*Wholesale & retail trade, repair of motor vehicles & motorcycles*; through H—*Transportation & storage*; I—*Accommodation & food service activities*; J—*Information and communication*; K—*Financial & insurance activities*; L—*Real estate activities*; M—*Professional, scientific & technical activities*; N—*Administrative & support service activities*; O—*Public administration and defence, compulsory social security*; P—*Education*; Q—*Human health & social work activities*; R—*Arts, entertainment & recreation;* to S—*Other service activities*. The sections feature numerous subcategories, but already a diversity of activities is clear.

Seeing services as "doing things" leads us to think about the transformations that are effected by different service industries. For example, sections G and I mainly involve making physical goods and buildings available to, and in a fit state for, people (in shops, restaurants, hotels, garages, etc.). Sections J, K, L M and O involve the processing and communication of information (taking the money and property rights dealt with in sections K and L to be a matter of information); N consists mainly of informational activity, too, of a less knowledge-intensive form; but the statistical classification also places some more physical activities like office cleaning, and packaging services into this section. Services mainly oriented toward transformations of the state of people are harder to disentangle, because these transformation may be physical (movement in space), biological (e.g. surgery, medical treatment), cognitive (education and training) or affective (entertainment). Section Q is most clearly a matter of transforming the state of human beings; to some extent O is also so engaged, while much of the service activity of sections P and R is transforming the state of information and of its users. (Of course, some production and delivery of information features in just about all service activities— and most of them ultimately involve changes in the state of people.) Section H involves transport of both goods and people, which sometimes involves the same vehicles and infrastructure, but which can involve quite separate systems.

The focus on the transformations that service industries effect can readily be used to examine the innovation trajectories of different types of services—which has bearing on the future of services, as well as on the past. Motor power, and petroleum and electrical energy in particular, have been applied to **physical transformations** such as those involved in providing transport, cleaning, etc. (Future patterns of innovation and system design are liable to be shaped by challenges of energy efficiency and reduction of CO_2 emissions.) Service industries have adopted new technologies, but so, as the classic Gershuny account stressed, have consumers. In some cases this means that consumers have been producing self-services with their own appliances, effectively in competition with such physical services as laundries and public transport. One result has been a decline in some of these services in terms of shares of employment and value-added. This has not been universal, however. Changes in lifestyle (such as female employment) and consumer taste, possibly coupled with organisational and product innovation on the part of service suppliers, have meant that restaurant and especially fast food service industries have often been rather buoyant. In part this may also reflect the ways in which these latter services can be a matter of entertainment and experience, and not just alternatives ways of accessing food and drink.

Service activities centred on **transforming information** have been the sites of particularly intense innovation connected with new IT, of course. Innovation has followed the evolution from mainframe computers and analogue communications through minicomputers, personal computers and the present era of portable devices, tablets and mobile digital communications. Many types of information previously distributed through print publications and a variety of recording media are now increasingly delivered online in intangible forms, and a host of new services have emerged to capitalise on, for example, the scope for using locational data to provide users with maps and information on what people and services are available in their vicinity. As devices that monitor human health and wellbeing become more widely available, we can anticipate much more development of information services addressing these issues. These aspects of the service activities more focused on transformations of the state of human beings are liable to be one of the major areas of service development in coming decades.[13] There is also considerable scope for development of new services processing and supplying information to support leisure and social life, as well as the business processes of organisations of all types.

New IT is important for **human transformations**, both because there is often much communication with the human recipient of the service, and because only with high levels of data storage and processing power can the diversity of human individuals and their needs be taken into account. But other sorts of technological and organisational innovation are also appropriate, ranging from the pharmaceuticals and prosthetics of health services; through the "scripts" and emotional sensitivity required for interventions like cognitive-behavioural therapy, mindfulness training, massage and physiotherapy; to the organisation of servicescapes, rides, and other experiential aspects of theme parks and other leisure facilities. Many different lines of development have emerged in the past around different aspects of human characteristics and practices, experience and wellbeing. Health services in particular are liable to continue to be reshaped by technological developments such as gene therapies, pharmacogenomics, and the like, and organisational developments integrating health and social care. Some common underlying knowledge bases are liable to emerge, however, that can be used to underpin many innovations concerning human learning and social interactions in the next decades: advances in neuroscience are one source of such knowledge.

Thinking about the range of transformations effected by services provides some insights into past and future patterns of innovation across different types of service activity. Loosely related to these transformations are varieties in the production of services. Most traditional services, and much of the service management literature, have concerned **Human-to-Human** (H2H) services, where the service interaction is largely between the customer and a human service supplier (or with several employees of a service firm). H2H service systems inevitably involve more than just this interaction between humans—their architecture also involves a "servicescape" of dedicated buildings and physical infrastructure, or support by material tools (such as surgical, teaching, restaurant, and transport equipment). Increasingly, service suppliers have moved to formats that link **Humans** with **IT systems**. In these H2IT (and IT2H) formats, people interact with and acquire services from terminals, websites and other IT agents and interfaces—whether or not human beings are involved at some point in approving, packing, dispatching, or delivering the core service. We may anticipate automated equipment[14] and robotics emerging with lower costs and greater functionality, and finding application in

---

[13] For one of the most important lines of development here—personal health systems—see Pombo-Juárez et al. (2014). New IT is also being used in many other applications, such as robotic surgery, screening the genetic structure of diseases and patients, and 3-D imaging (as in the case of my new tooth!).

[14] E.g. driverless vehicles; these have long been familiar in some rail services and warehouses, but are a novelty in other settings.

some services. (Unpleasant and hazardous jobs may be displaced, for example—but so might others, with possible threats to low-wage employment.) IT systems interact in **IT system-to-IT system** frameworks—IT2IT—famously in "automated trading" in financial services, where concerns have been raised about lack of human oversight and dangers of financial volatility.[15] In everyday life, we have become accustomed to PCs, smartphones and "cable boxes" updating software from network providers; to search engines automatically updating newsfeeds or other information requests; to software agents bidding on auction sites like eBay, and to our in-car navigation systems tracking signals from satellites.

It is tempting to see the main trend as being from H2H to H2IT and IT2IT services, and there is much talk these days of the scope for robots to accomplish tasks that are currently performed by people - more-or-less humanoid robots serving customers in restaurants and helping to care for sick and disabled people, more specialised devices performing brain surgery and providing automated transport. As IT power increases and costs fall, such options are presumably going to be pursued. But there are counterforces (for example, where "high-touch" is valued more than "high-tech")[16]—and innovation can also produce new H2H services (as in the case of cognitive-behavioural therapy, mentioned above, which compares well with technological solutions like psychotropic drugs).[17] One of the vagaries of our current innovation policies is the ways in which they support technological innovation (through research programmes, tax credits for R&D, and the like) but largely neglect organisational and other elements of service innovation.

Service industries and organisations also vary dramatically in terms of their workforce's skills and knowledge. Some service industries have huge shares of highly qualified people in their staff—KIBS, education, health services, for example—compared to the rest of the economy. In contrast, some feature outstandingly large workforce shares of people with very low educational attainment (especially in McJobs and in sectors like hotels and catering). Yet others employ large shares of "symbol processors" undertaking information work for large organisations (e.g. public administration), with clerical and secretarial skills in particular; the first waves of office automation seem to have impacted particularly heavily on such middle-range jobs in larger organisations. "Flatter" organisations have used new IT to reduce the number of hierarchical steps required to process, centralise and distribute information.[18] Additionally, many of the more routine informational tasks have been outsourced and offshored to cheaper locations, along with much routine manufacturing work. There is considerable debate as to how far more

---

[15] See Government Office for Science (2011) for a forward-looking study of this topic.

[16] Toffler (1980) introduced this couplet, along with other relevant ideas (such as "prosumer"). For an interesting education rendition of this, see Moursound (2004).

[17] Reviews of the effectiveness of CBT can be found at, for example, http://summaries.cochrane.org/search/site/cognitive%20behavioural%20therapy. As might have been predicted, IT is being brought into play here, too, with computerised CBT being rolled out.

[18] The work of David Autor is particularly relevant here—see Autor and Dorn (2011), for example. Some European evidence on polarisation is provided by Eurofound (2013).

professional service activities can be conducted at a distance in this way—while some tasks, such as software writing or preparing fairly routine accounts may be fairly easy to offshore, others require more H2H contact. The extent (and timescale) to which this can be accomplished through improved IT systems remains to be seen.

It remains controversial as to how far these developments are responsible for the polarisation of wages and employment opportunities that many Western countries have witnessed in recent years, though they are most probably part of the explanation. Inequality is currently high on the political agenda in many countries; the long-term implications for social mobility of the disappearance of many of the steps in the traditional career ladder is concerning for the future. The future of the services economy will look very different if we move even further into a polarised society (walled enclaves, expanding markets for security services, some combination of high tech bread and circuses and pacification for the masses) than if we are able to recreate some version of the postwar meritocratic consensus that prevailed (at least in rhetoric, if less consistently in practice) in most Western societies.

14.4   The Future of the Service Vista

Foresight activity is bound to require understanding of the future prospects for service thinking and services activities. Work specifically on the future of services could fruitfully combine (1) analysis of specific industries and activities, then, with exploration of the opportunities for innovation to do new things, and to do things in new ways, within these areas, with (2) examination—perhaps involving scenario analysis—of broader trends and uncertainties in the development of occupational structures, markets and the broader political economy. In this context it will be worthwhile paying particular attention to the factors that lie behind the growth of business services, and KIBS in particular; to the reshaping of consumer markets as social trends such as ageing populations and migration patterns continue to evolve; and the challenges associated with the changing role of the state where it comes to public service provision and the regulation of many service markets. It will be necessary to see service systems in an international context, with new service models being imitated across countries and with service transnationals active in many locations (and pushing for more access to national markets through trade agreements). The broader context of "Grand Challenges" to the environment, human health and security, resource depletion, and the like, also needs to be confronted.

This latter point is integral to thinking about the future of services. For whether our Grand Challenges are attributed to the success of agricultural and manufacturing industries in producing an abundance of material goods (at some cost to the environment) in some parts of the world, or as reflecting structural problems in our social organisation, it is fairly clear that these are "wicked problems" which brook of no easy solutions.[19] They involve multiple factors (and

---

[19] The term was introduced by Rittel and Webber (1973) and a large literature has grown up around it, especially in policy studies. A helpful useful review is Australian Public Service Commission (2007).

stakeholders, many of whom have different points of view and understandings of the challenge), there are numerous feedback loops between these, and often the effective and sustainable ways of responding to them require behaviour change (and ways of motivating behaviour change). There are usually both social and behavioural changes required, in fact, and knowledge of the ways of effecting and coordinating these requires knowledge to be shared and fused across different fields of professional expertise and jurisdictional/departmental authority.

A Service Perspective may prove helpful in addressing many of these grand challenges. This is not just because service organisations (for example, land and water remediation and waste processing; care of elderly and disabled people; disaster relief; and services concerned with energy management, eco-auditing, and so on) are already involved in the current and partial responses to grand challenges. It is because we can think of the challenges as involving service systems—because they are about threats to the creation and delivery of all sorts of vital services to large populations. Three ideas can be mentioned in this context.

First, as the Service Dominant Logic argues, all economic activity can be seen as ultimately a matter of providing service to somebody. (Note that the service process may involve transforming people who are not actually the paying customers of the service—prisoners are held securely to provide wider social benefits, not because they request incarceration; we are all targeted by the products of the advertising industry, though sometimes the trade-off is that we get access to free newspaper, website and TV services.) Often many services are produced and consumed in the course of providing the services ultimately required by the population: design and engineering services, financial and transactional services, security and transport services, and many more, are all involved in the supply of transport, retail, health and just about all services that we ultimately consume.

Second, and related to this, we can see the economy as composed of a complex of interacting "Product-Service Systems", in which goods and services are being produced and consumed in order to support the main functions of societies. Using the notion of Product-Service Systems in order to consider how Grand Challenges may be confronted is an approach that has most frequently been invoked in order to examine whether material resource consumption could be reduced by alternative modes of service provision—for example, instead of every individual purchasing a car, systems of car hire, pooling and sharing might be employed. (The difficulties encountered by such alternative system organisation are suggestive of how wicked problems are not soluble by simple solutions or "magic bullets".) Service Systems are viewed as consisting of POTI (people, technology, organizations, and shared information) organised in value coproduction networks.[20] But more recently methods for analysing and (re)designing Service Systems have attracted much attention. In large part this reflects the emergence of the SSME (Service Science,

---

[20] Among the publications dealing with the SSME approach are: Demirkan et al. (2008), Hefley and Murphy (2008), Maglio et al. (2010), Maglio and Spohrer (2008), Spohrer et al. (2007). A Delphi study of implications for curriculum design is Choudaha (2008).

Management and Engineering—or is it Education?) approach. This approach represents major initiatives from (mainly) IT-based corporations (especially IBM), who determined that more systematic appraisals of their business and the architecture of their products was in order: it was no longer enough to focus on selling hardware.

Together, these two sets of ideas suggest that more attention to service systems, and the services used in and yielded by economic organisation, and development of methods for design and redesign of such systems (and for testing and scaling up designs), will be an important part of tackling Grand Challenges into the future. A third idea also bears on this.

This idea—which does not sit easily with the Service Dominant Logic approach—is that services are being supplied to humanity by the natural world. The idea of "eco-system services", such as the way in which (for example) forests help supply us with fresh air, reduce flood risks, promote biodiversity, has gained considerable traction (Wallace 2007) and attracted much controversy (e.g. Potschin and Haines-Young 2011). Though people may sometimes seek to attach monetary values to these services (not least in order to provide an economic argument about the costs of environmental damage), it is evident that the Earth is not offering us services as part of a transaction. But then again, many H2H services are not transactional in an economic sense. Parents' care for their children, passers-by aiding victims of road accidents, even politicians and public servants spending their time in support of worthy causes (yes, some do!) can also be seen as providing services. In other words, value can be created without being valued in monetary terms, and without money changing hands.

Recognition of this is, of course, important for environmental conservation—the natural world is not just a source of beauty, but also of life-support systems. But it also brings to the fore that our social systems extend well beyond the economic sphere, and it will be important to bear this in mind when thinking about service systems and their future. Innovation may not just be a matter of developing new things and ways of doing things, in the formal economy; it can involve the wider society. One indication that awareness of this is growing is the renewed attention given to "social innovation"—a term that has been applied to many phenomena, but which we can see as especially relevant to innovations stemming from—and involving new roles for—third sector (grassroots and voluntary) organizations (see Mulgan et al. 2007; Murray et al. 2010). The future of services is not just a matter of big service organisations instituting new technologies—it also involves initiatives from across society oriented towards tackling problems and grand challenges, redesigning service systems for better and more inclusive lifestyles. Or so we can hope, and task policymakers with finding ways of promoting such a future.

This in no way reduces the need for new tools and new design approaches where it comes to service (system) development. Indeed, service design is liable to be critical for the future of service(s)—and the human future more generally. So let us express another hope: that we are on the threshold of an era where a wide range of compelling and participatory service design tools are coming into wide and successful use.[21]

Acknowledgement The article was prepared within the framework of the Basic Research Programme at the National Research University Higher School of Economics (HSE) and supported within the framework of the subsidy granted to the HSE by the Government of the Russian Federation for the implementation of the Global Competitiveness Programme.

References

Australian Public Service Commission (2007) Tackling wicked problems: a public policy perspective. Government of Australia, Canberra

Autor D, Dorn D (2011) The growth of low-skill service jobs and the polarization of the US labor market. Am Econ Rev 103(5):1553–1597

Barras R (1990) Interactive innovation in financial and business services: the vanguard of the service revolution. Res Pol 19(3):215–238

Bettencourt L, Ostrom A, Brown S, Roundtree R (2002) Client co-production in knowledge-intensive business services. CA Manage Rev 44(4):100–128

Bitner MJ (1992) Services capes: the impact of physical surroundings on customers and employees. J Market 56(2):57–71

Choudaha R (2008) Competency-based curriculum for a Master's program in service science, management and engineering (SSME): an online Delphi study. PhD thesis: Morgridge College of Education, University of Denver

Dachs B, Biege S, Borowiecki M, Lay G, Jäger A, Schartinger D (2012) The servitization of European manufacturing industries. Austrian Institute of Technology, Vienna

Demirkan H, Spohrer JC, Krishna V (2008) The science of service systems. Heidelberg

Doroshenko M, Miles I, Vinogradov D (2014) Knowledge-intensive business services: the Russian experience. Foresight-Russia 8(4):24–38

Eiglier P, Langeard E (1987) Servuction. McGraw-Hill, Paris

Eurofound (2013) Eurofound yearbook 2013: living and working in Europe. European Foundation for the Improvement of Living and Working Conditions, Dublin. http://www.eurofound.europa.eu/publications/htmlfiles/ef1416.htm. Accessed 7 August 2015

Eurostat (2009) Statistical classification of economic activities in the European Community, NACE Rev 2. Eurostat, Luxembourg

Evangelista R, Savona M (1998) Patterns of innovation in services: the results of Italian innovation survey. In: Paper presented at the 7th Annual RESER Conference, Berlin, October 8–10

Gartner A, Reissman F (1974) The service society and the new consumer vanguard. Harper and Row, New York

Gershuny JI (1978) After industrial society? Macmillan, London

Gershuny JI (2000) Changing times: work and leisure in postindustrial society. Oxford University Press, Oxford

Gershuny J, Miles I (1983) The new service economy. Frances Pinter, London

Government Office for Science (2011) Future of computer trading in financial markets: working paper. UK Foresight Project on the Future of Computer Trading, London.

---

[21] For discussions and practical experience in the emerging fields of service design, see the websites of the service design network (http://www.service-design-network.org/) and the service design research network (http://www.servicedesignresearch.com). The journal *Touchpoint* focuses on this area.

https://www.gov.uk/government/publications/future-of-computer-trading-in-financial-markets-working-paper. Accessed 7 August 2015

Gould AM (2010) Working at McDonalds: some redeeming features of McJobs. Work Employ Soc 24(4):780–802

Hefley B, Murphy W (eds) (2008) Service science, management and engineering education for the 21st Century. Heidelberg

Hipp C, Tether B, Miles I (2000) The incidence and effects of innovation in services: evidence from Germany. Int J Innov Manag 4(4):417–454

Hortelano MDE, Gongález-Moreno A (2007) Innovation in service firms: exploratory analysis of innovation patterns. Management Research: J Iberoamerican Acad Manag 5(2):113–126

Levitt T (1976) The industrialisation of service. Harvard Bus Rev 54(5):63–74

Miles I (1987) The convergent economy. (Imperial College/SPRU/TCC series: Papers in Science, Technology and Public Policy.) Imperial College, London

Maglio PP, Kieliszewski CA, Spohrer JC (eds) (2010) The handbook of service science. Heidelberg

Maglio P, Spohrer J (2008) Fundamentals of service science. J Acad Market Sci 36(1):18–20

Miles I (1993) Services in the new industrial economy. Futures 25(6):653–672

Moursound D (2004) High tech/high touch: a computer education leadership development workshop. ICCE, Eugene. http://pages.uoregon.edu/moursund/Books/HT-HT/HT-HT.pdf. Accessed 7 August 2015

Mulgan G, Ali R, Halkett R, Sanders B (2007) In and out of sync: the challenge of growing social innovations. NESTA, London (see also the text available at: http://eureka.bodleian.ox.ac.uk/761/)

Murray R, Caulier-Grice J, Mulgan G (2010) The open book of social innovation. NESTA, London. http://www.nesta.org.uk/publications/open-book-social-innovation. Accessed 7 August 2015

Pombo-Juárez L et al. (2014) Personal health systems: foresight synthesis. Austrian Institute of Technology, The University of Manchester, Impetu Solutions. http://www.phsforesight.eu/reports. Accessed 7 August 2015

Potschin MB, Haines-Young RH (2011) Ecosystem services: exploring a geographical perspective. Prog Phys Geogr 35(5):575–594

Rittel HWJ, Webber MM (1973) Dilemmas in a general theory of planning. Pol Sci 4(2):155–169

Spohrer JC, Maglio PP, Bailey J, Gruhl D (2007) Steps toward a science of service systems. IEEE Computer 40(1):71–77

Sundbo J (2002) The service economy: standardisation or customisation? Serv Ind J 22(4):93–116

Teichgraber M (2013) European Union labour force survey—annual results 2012. Statistics in focus 14/2013 http://ec.europa.eu/eurostat/statistics-explained/index.php/Archive:Labour_force_survey_overview_2012. Accessed 7 August 2015

Toffler A (1980) The third wave. Bantam Books, New York

Vargo S, Lusch RF (2006) Service-dominant logic: what it is, what it is not, what it might be. In: Lusch RF, Vargo S (eds) The service-dominant logic of marketing: dialog, debate, and directions. M.E. Sharpe, Armonk

Wallace KJ (2007) Classification of ecosystem services: problems and solutions. Biol Conserv 139 (3–4):235–246

Future-Oriented Positioning of Knowledge Intensive Local Networks in Global Value Chains: The Case of Turkey

15

Erkan Erdil and Hadi Tolga Göksidan

15.1 Introduction

In the current literature of global economics research, we may depict different insights for competing in a global value chain (GVC) which may build up a foundation for the industrial innovation and learning (e.g. Gereffi 1994 and 1999). Eventually, we may also list many ways to achieve to build up this foundation. First, we may underline "process innovation" as a tool to improve the efficiency of transforming inputs into outputs. Only by this way, the internal processes become significantly better than those of rivals, both within links in the chain (more inventory turnovers, less scrap) and between links (more frequent, smaller and on-time deliveries). Second, we may underline "product innovation" as a leading tool to achieve better quality, lower priced and more differentiated products, as well as shorter times to market for new products. Third, we may underline "functional innovation" as a tool to achieve new responsibilities for new activities in the GVC. As a forth, "inter chain innovation" helps enterprises to move into new and more profitable chains.

In developing countries, like Turkey, some enterprises may even latch onto several GVCs, providing further opportunities for linking to local enterprises connected with them. Such SMEs lift themselves—and those connected with them in supply chains—to new levels of performance and quality, driving forward the momentum of collective industrial development.

This article depicts some important effects of GVCs on developing countries as it helps on shifting links and contractual relations among transnational companies

E. Erdil (✉) • H.T. Göksidan
Department of Economics, Science and Technology Policies Research Center, Middle East Technical University, Dumlupınar Bulvari, 06800 Ankara, Turkey
e-mail: erdil@metu.edu.tr; htolga@gmail.com

© Springer International Publishing Switzerland 2016
L. Gokhberg et al. (eds.), *Deploying Foresight for Policy and Strategy Makers*,
Science, Technology and Innovation Studies, DOI 10.1007/978-3-319-25628-3_15

245

and SMEs. Hereby, we expect enterprises to expand their product lines, and to expand internationally by forging new links with enterprises already active in the global economy, encompassing research and development, production, logistics, marketing and exchange, where all the links are between enterprises rather than between countries.

In fact, developing countries provide a means for accelerating the development of enterprises and countries, providing openings that developing country enterprises can exploit to upgrade their capabilities. For such enterprises, or local clusters of enterprises, the task is to insert themselves into the wider networks. This may be regarded as the main achievement for sustaining competitiveness, in similarities with the re-structuring of regional networks in developing countries that often compete by participating in extensive inter-firm networks.

As another dimension in our study, we will investigate and argue whether if it is possible to increase and improve the participation of Turkish's SMEs in the global economy, which is explicitly the baseline hypothesis of this study. The literature on regional networks and GVC, which are mainly focused on analyzing the local sources of competitiveness from vertical and horizontal intra-cluster relationships that generate collective efficiency, has barely investigated the increasing importance of external international linkages. Hence, this study will provide some new insights to show the international linkages of Turkish SMEs, which often lack the capabilities to participate effectively in global markets (e.g. Peres and Stumpo 2000, 2002). The following question is central to this study: What can be done to support SMEs' global market linkages regarding the Turkish regional networks?

In developing countries like Turkey, the GVC analysis has shown recently how international linkages can play a crucial role in accessing technological knowledge and enhancing learning and innovation (Altenburg 2006; Gereffi 1994, 1999; Gereffi and Kaplinsky 2001; Giuliani 2005; Kaplinsky 2000; Humphrey and Schmitz 2002a, b; Pietrobelli and Rabellotti 2007).

According to Morrison et al. (2008), value chain research focuses explicitly on the nature of the relationships among the various actors involved in the chain, stressing the role that global buyers and producers may play in supporting developing countries producers' learning and innovation activities, and explores their implications for development. In this respect, the concept of networks among suppliers and buyers is central to this analysis. Related to this contribution, here, we must denote that there has been numerous approaches to favor vertical linkages, knowledge transfer and productivity spillovers among the networks of domestic and foreign firms. With a lesser degree of research on the issue of GVC, in literature, researchers have significantly drawn attention to the variety of value chain relationships wherein global buyers interact with local suppliers in different countries. Saliola and Zanfei (2009) denotes that alternative relationships (governance modes) will emerge in the presence of different degrees of standardization of products and processes, and of different competencies of suppliers.

Hence, as a complementary approach, the aim of this study will, indeed, explore if and how GVC structure fosters knowledge transfer and innovation in developing countries as in the case of Turkey. The general literature which will be presented on

GVCs (Gereffi 1999; Gereffi and Kaplinsky 2001) draws attention to the opportunities for local producers to learn from global leaders (buyers or producers) of the chains within different mechanisms of knowledge transfer. This study will construct the scope and pattern of regional networks that facilitates the creation of global linkages in a Turkish SMEs and MNCs perspective. Finally, the study addresses the following specific questions with regard to the specific Turkish case:

1. Are SMEs' global linkages facilitated by the degree of regional networks?
2. How do regional networks embedded to GVCs in such a way that supports organizational learning and strengthens the linkages among SMEs?

15.2   Basic Definitions and Notations on the Theory and Application: International Production Networks and Global Value Chains

The shift in the structure of international trade poses challenges to both economic theory and policy. The challenge here is to cope with the rise of international capital mobility and trade in intermediate goods with regard to international trade and foreign direct investment. Hence, by sustaining a relative advantage that gives way to compete in global markets, the relative decision making for a (part) of production process; even with respect to foreign investment; highly depend on the interpretation of the application of some externalization theories, simultaneously creating the need for an economic theory of internalization.

In fact, GVC provides two insights about innovation and trade. First, creating value is not confined to only production. In relative advantage that GVCs create, products are brought to market through a combination of activities of transnational companies. By this way, we may argue that enterprises can succeed in improving capabilities in production, developing new capabilities outside production (design and marketing skills), diversifying customers and market destinations, developing the capacity to introduce new products or to imitate leading innovators quickly and successfully.

Besides, as the most important fact, we must denote that the advantage of GVCs is that enterprises can seek involvement at their level of technological competence. For instance, in Turkey, most of the enterprises were vertically integrated in supplier networks that did not offer much scope for skills enhancement and innovation. Hence, the globalization of production comprises both international trade and foreign direct investment with great promise of a new phase of export growth from developing countries whose inclusion in the process opens new markets and introduces new technologies for the enterprises. Moreover, as world trade has expanded, one can assume that the developed nations have fostered their share of services to developing countries (see Table 15.1) while developing countries have increased their share of manufactured goods to developed nations.

Table 15.1 Exports of goods and services as a share of total exports from developed and developing countries

	Percentage (%)			
	Developed countries		Developing countries	
	Exports of goods as a percentage of total exports	Exports of services as a percentage of total exports	Exports of goods as a percentage of total exports	Exports of services as a percentage of total exports
1998	78.3	21.7	83.2	16.8
2012	76.1	23.9	85.9	14.1

Source: DPAD calculations based on IMF Financial Statistics

This tendency points out that, since the late-1990s, there has been a major rise in the share of developed countries in services exports, and the decline of the share of manufactured goods exported to developing countries. Here, we can say that the next stage on that the pace of globalization of production may come with a great promise of a of services export growth from developing countries, whose inclusion in the process opens new markets and introduces new technologies. Furthermore, as world trade has expanded, both in absolute terms and in relation to world output, developing countries have maintained their share of world exports of manufactured goods, while the internalization of production operations have induced the development of asset accounts through foreign direct investment (FDI) as a result of sustaining such internal knowledge assets that enables firms to invest abroad.

Moreover, as an old-established theory and concept in the economics literature, the value chain or value-adding chain has been used most prominently by Porter (1990) and has achieved very wide acceptance in the management community (Henderson et al. 2002). Simultaneously, as pointed out by Henderson et al. (2002), the emphasis is intensely on the sequential and interconnected structures of economic activities (like the analysis of different levels of FDI targeted at a different sector) with each link or element in the chain adding value to the process of production networks. Here, it can be denoted that Porter's study may just be considered to be a partial analysis since it is bounded by the firm or inter-firm networks and is barely explaining the effects of the institutional contexts of firm-based activities, or the formation of vertical relations in the embedded network forms.

In this regard, to understand the full scheme of the global dynamics of this progress, one also has to focus on the role of local linkages in generating competitive advantage in developing countries. The sectoral and local scheme on this economic research topic are termed to be exactly the ones in which global buyers (whether agents, retailers or brand-name companies) have come to play an increasingly important role in the organization of global production and distribution systems. Here, once again, we would like to mention that as one of the main literature which analyses these global systems, GVC research is a different approach whether to the question of upgrading, emphasizing cross-border linkages between firms in global production and distribution systems rather than local linkages (see Gereffi and Korzeniewicz 1994; Gereffi and Kaplinsky 2001).

Since the GVC approach is weak in explaining local upgrading strategies, in order to solve this dichotomy, one must distinguish between different types of local networks and different types of chains (Humphrey and Schmitz 2004). Here, through vertical integration among firms, the formation of GVC in a local manner must be exemplified in two respects; local networks bringing together partners with complementary competences, and vertical relations in which the innovation capability and competence levels were leveraged in favor of the global buyers. We will further investigate whether the governance forms of GVC coincides the upgrading of local firms, explaining why it is important in the case of developing countries.

Very briefly, the extensive work by Gary Gereffi conceptualizes the chain of economic activities as a global commodity chain (GCC).[1] In his work, the characteristics of the GCC framework have been extensively outlined as:

> ... sets of inter-organizational networks clustered around one commodity or product, linking households, enterprises, and states to one another within the world-economy. These networks are situationally specific, socially constructed, and locally integrated, underscoring the social embeddedness of economic organization. (Gereffi 1994, p. 2).

Besides, GVC provide a means for accelerating the development of enterprises and countries by helping to exploit for upgrading their capabilities. For such enterprises, or local clusters of enterprises, the wider aim is to take place in the wider networks. This status requires an initial base of technological capability by default, and generously built upon some purposive innovation and collective learning.

Hence, in an aim to access to worldwide markets and to retain knowledge of other global players in the world economy, generally, SMEs of developing countries seek involvement at their level of technological competence. For example, in Turkey, machinery producers have weak vertical linkages in globally integrated supplier networks that furnish the required global skills to innovate. However, within the different trade agreements, buyer groups from Europe and Asia (including some big multinational companies) have started to create alternative global value chains that offer SMEs a greater scope for expanding their responsibilities for innovation. Here, as an example, this progress allows white good firms in Turkey to develop some certain capabilities up to higher levels in GVC.

Moreover, we might say that some firms are bounded (and embedded into) to several GVCs providing further opportunities for linking other local enterprises that are in any kind of economic relation with them. Such firms (in theory, they are focal firms) simply adopt themselves (and those connected with them in supply chains) to new levels of learning and innovation to achieve the goal of industrial development. As a well known economic and theoretical fact, such industrial learning is a long

---

[1] See Gereffi and Korzeniewicz (1994), Gereffi (1999) and other studies on GCC.

and strenuous process that in this ongoing process, the GVC offers spontaneous technological and economic structures to link local firms to global networks.

Nevertheless, if we aim to show that the GVC theory is ample to explain industrial development and innovation in developing countries in the context of increased globalization and transnational inter-firm linkages, one must give focus on the regional structures as with the processes of technological capability development and innovation on the firm-level and with the other contextual factors enhancing on the evolution of this process. The studies on technological capabilities in developing countries perspectives (see Lall 2001; Pietrobelli 1998) may also lead to clear understanding for the integration of the GVC literature and for building up an empirical framework to explain local industrial developments in developing countries. Drawing upon the evolutionary approach of Nelson and Winter (1982), the technological capabilities literature claims that technological change is the result of purposeful investments undertaken by firms, and therefore transfer and diffusion of knowledge and technology are effective only in so far as they also include elements of capability building.

Moreover, GVC literature can fully exploit the theories of innovation and knowledge in a developing context by explaining the different levels of networking and the degrees of knowledge transfer that affect the GVC governance structure, and the speed of learning on the role of local linkages in generating competitive advantages in export industries. Hence, in terms of the micro-level processes of knowledge transfer, learning and networking, we will issue a number of facts that need to be addressed in this effort. For example, in order to elaborate the theory of GVC, one of the most important facts is what occurs at the firm level, on the mechanisms of learning, networking and innovation, as proposed by the GVC approach by drawing attention to some regional development strategies focused on some key features of knowledge transfer.

In the following sections of this study, we will deal with the issue of new forms of international organization of more complex production processes arisen from the development of new knowledge-intensive local networks that certainly have brought us about a criticism to the concept of GVC as part of a complementary way of knowledge generation that are highly associated with theoretical economic changes and development in the local and global economy.

Within this context, we will integrate the concept of production networks from a methodological and theoretical perspective that is simply based on two dimensions. These dimensions are:

- The local and global supplier—buyer linkages among agents in a regional network theory perspective;
- Knowledge transfer and learning including organizational and institutional perspectives in a knowledge theory perspective.

15.3   Regional Networks: The Local Linkages Among Agents

Innovative firms are linked to the outside world by various kinds of connections, in particular, international linkages with customers and suppliers, as a key requirement for successful development of innovations (Doloreux and Parto 2005). Commonly, networks provide firms a wide range of knowledge sources that not only generates inputs for firms but also sustains their economic activity. Recent contributions by Bathelt (2003) and Malecki and Oinas (2000) among others, have pointed out the importance of local interaction and global connections for understanding the competitive advantages of innovative firms and regional clusters (Doloreux and Parto 2005).

The concept of regional innovation systems focuses on localized learning processes to sustain the competitive advantage of regions. In an aim to develop such policy measures, the regional innovation systems framework furnishes firms to develop certain capabilities as well as to improve their business environment. From this standpoint, it should be said that it is crucial to support the creation of interactions between different innovative actors such as between firms (supplier–buyer relations) and universities or research institutes, or between small start-up firms and larger (customer) firms (Doloreux and Parto 2005).

In industrial supplier–buyer relationships, buyers and suppliers together create core competencies in different industrial functioning states. It is also denotable that these competencies may also sustain continuous learning and differing levels of production efficiency. When these competence powers were combined in a network of firms, the networking advantage subsidizes firms to access to critical resources that enable the creation of superior value even in the international marketplace.

To further explain the empirical analysis of network formation and capabilities that influence performance, we propose that an important dimension on which firms differ is the extent of inter-firm (production network) specialization. The performance of a firm is directly related to which the firm and its suppliers make collaborative investments at all. In particular, we argue that firms may develop some certain competitive advantages when they try to participate in a production network characterized by a high degree of inter-firm specialization.

Regarding a brief outlook of historical background of economics and the formation of the production networks in Turkey, we can say that the Turkish national policies related to industrial development locations are stimulating the formation of agglomerations of similar-sector firms. Due to basic networking concerns, SMEs in the manufacturing sector are encouraged to locate in the appropriately planned "small industrial estates" and "organized industry zones". These places are planned and managed according to different regulations and incentive methods to encourage appropriate firms to locate and operate in these areas. The basic aim in developing this type of formation in regions is to provide firms with an effective business environment that contributes their competitiveness and eliminates the drawbacks related to infrastructure, bureaucracy etc.

As these locations are the places of agglomeration of firms, they form an environment that the clusters are likely to emerge (or exist) in by market-induced mechanisms related to Marshallian aspects of the study (Özcan 1995). Therefore, SMEs in the Turkish Economy are attributed great importance and various technological and financial instruments have been developed for the provision of support (Eraydın and Armatli-Köroğlu 2005). Since 1996, which was announced as SMEs year in Turkey, the situation of SMEs in Turkey has been handled by strong attention. The importance of SMEs in addressing the triple challenge of more growth, greater competitiveness, and more jobs has been brought into ever-sharper focus over the past few years (Kuruüzüm 1998). Also, the necessity of effective integration of the Turkish SMEs to international economic area also stresses the importance of SME support policies and the need for an effective GVC approach to increase the competitiveness of the Turkish SMEs to compete globally. Unfortunately, one can say while various public policy instruments are employed to support Turkish SMEs, still, the desired level of competitiveness has yet not achieved (Kuruüzüm 1998).

Moreover, Eraydin and Armatli (2005) depict that the industrial agglomerations, which are denoted as "Turkish production networks" in this paper, are formed to be an outcome of the economic and spatial transformation that has been taking place in Turkey since the beginning of 1980s. In fact, according to the authors, the 1980s became the turning point of economic policies in Turkey, from protectionist attitudes which dominated Turkish national economic policy prior to this period to increasing reliance on market forces. While the new program greatly freed up foreign trade and exchange, in 1984 major structural changes further liberalized trade by dismantling foreign exchange controls and quotas on imports, and by revising tariffs. The liberalization initiative has continued by export promotion policies, the depreciation of exchange rates and direct subsidies. The efforts of economic transformation are further supported by several private, semi-public and public institutions. Regionally, the economic transformations, the new competitive environment and the loss of protectionist policies also enforced the spatial transformation in Turkey. While the areas with relatively developed manufacturing capacities became the cores of export activities, hence, the regions with a weak manufacturing basis had obvious difficulties in becoming linked to the newly-organizing international production networks.

In this respect, a pioneering attempt to identify and analyze industry clusters in Turkey is done in the context of the "Competitive Advantage of Turkey" project, in association and consultancy with the Centre for Middle East Competitive Strategy (Akgüngor 2003). This project aimed at analyzing the regional concentrations of industries at the mega-level cluster and network analysis applications. The attempts focused on identifying national cluster templates by examining buyer-seller relationships across industries through input–output based analysis. By referring this project, the complementary study by Akgüngor (2003) was to interpret the on-going project results aiming to investigate further regional

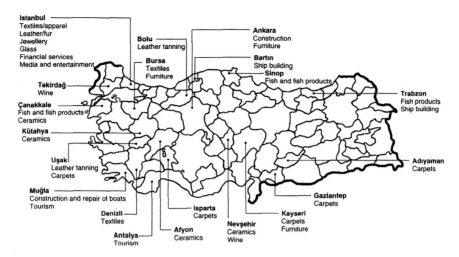

Fig. 15.1 Selected examples of highly concentrated industries in Turkey. Source: Öz (2004)

concentrations of cluster templates and to identify high-point industries within the identified regional clusters. Moreover, in the study, classification of the clusters according to their potential for decline or growth in each of the geographical regions of Turkey is provided. While these initial studies provide valuable policy information for the regional development efforts, as Akgüngor (2003) notes, the research should be expanded in order to explore the clusters at the micro level and further explore formal and informal ties across the industries and institutions.

What has been striking during this spatial economic transformation is the increasing importance of some industrial agglomerations that are located far from the earlier manufacturing cores, in terms of production and exports. Obviously, these new industrial agglomerations are located in different parts of the country (see Fig. 15.1) and achieved different stages of evolution.

The regions designated in Fig. 15.1 (and more studies) have been prepared especially on the areas of Denizli (Eraydin 1998; Erendil 1998; Armatli-Köroğlu and Beyhan 2003; Öz 2004), Bursa (Reyhan 1990; Ersoy 1993; Saraçoğlu 1993) and recently on Ankara (Tekeli 1994; Dede 1999; Erdil and Göksidan 2006). As an another focus in this manner, although the clusters considered in Eraydin and Armatli's (2005) work do not represent idealized industrial districts (or networks); parallel to authors' determinations; the basic characteristics of each production region and network clearly shows that each formation have different features, and furthermore, will help us to discuss further how productions networks can be supported under different structural aspects of business relations that are strictly important in the formation of networking patterns of relations for a firm located in these areas (see Table 15.2).

Table 15.2 Structural characteristics of selected clusters in Turkey: Denizli, Bursa and Ankara

Factors/ Conditions	Denizli	Bursa	Ankara
Type of the manufacturing cluster	• Industrial district	• Innovative manufacturing cluster	• High-tech industrial cluster
Area of specialization	• Textiles, especially towels and bathrobes	• Textiles for home furnishing	• Machinery, electronics, the defence industry and software
Main character of the cluster	• Traditional: small artisanal, and highly specialized family owned firms located in close proximity	• Traditional/Modern: small artisanal, and highly specialized firms as well as large multinational companies co-operating with these small enterprises	• Modern/high-tech: high-tech firms of different size
Main observed benefit	• Co-operation in production and marketing for international markets	• Collective competition in specialized fields	• Weak collaborative environment Market relations with state institutions
Technical dynamic Social capital	• Complementarities collaborative action, trust and reciprocity Strong social networks	• Specialization increasing shares of export in engineering industries, Adaptation and product development for international markets	• Adaptation of new technologies for national market Access to qualified labor

Source: Eraydin and Armatli-Köroğlu (2005)

Our view is that such network formation among firms and their suppliers involve more complex issues. In this study, we may argue that the Turkish subcontracting supplier—buyer relationships can be portrayed to have three main characteristics.

First, some of the networked relationships are *long-term* and duration is determined by the product-life cycles. Each time a new product is designed and manufactured, the large firm makes a call for the best offer from suppliers. At that stage, suppliers are put into competition. However, the firm generally continues subcontracting relationships with the same suppliers from a product to another, so that the firms can not solely be affected because of costly and timely renegotiations. Such duration of relationships allows deriving some of the benefits of *vertical integration*.

Second, some of the Turkish networked relationships are *institutionalized* and *hierarchical*. Such hierarchy of subcontractors is defined according to the type of product bought by the large firm. In this case, the subcontractors are autonomously chosen on the basis of quality. We must also mention that the design can also be jointly designated by the supplier and the firm itself. In the latter case, the supplier only executes orders from the firms according to its production definitions, and is highly dependent on the large firm.

Third, the Turkish networked relationships are *contractual* and *characterized by specific procedures*. The generic process (favoring innovation at all) is such that a contractual supplier is agreed, right before the new product is still in the development phase (with no specification of quantities to be delivered, nor the prices, etc.) providing flexibility and adaptation capability to possible changes in the specification of products at any time.

Therefore, the relationship between the networked firm and its suppliers can be characterized by the coexistence of cooperation and competition. Here, competition among rivals and other actors in the network prevails in the suppliers' selection phase, but also after the contract has been signed. Hence, we can say that the performance of suppliers in terms of quality and costs is indeed assessed and compared with other suppliers in the network. If the supplier does not perform well, orders are reduced and, in the last resort, the supplier is supposed to be changed. However, the firm has also an interest in cooperating with the supplier to avoid switching and associated costs (time to learn the specification of the product and production, time required to set up trust, etc.) which is a very typical case in the Turkish manufacturing industry. Furthermore, over time, suppliers are expected to share sensitive strategic data on a timely basis. This is the point where trust is needed. When suppliers and customers share information about their R&D expenditures, it encourages the supplier to invest in a customer's future needs. In Turkey, such contractual mechanisms do rarely work but this is especially critical when suppliers need to contribute on new processes and share tacit knowledge to make an investment in a new technology.

Consequently, manufacturers in Turkey seek suppliers who can help them to sustain their own product design capability and managerial skills, in order to continuously collaborate with, helping to resolve problems and exchanging continuously information in order to improve the system (Ulusoy 2003). The know-how generated by such a relationship is, according to Asanuma (1989), twofold. On the one hand, it is technical, regarding the product and production system. On the other hand, it is "relational", due to the incentives and knowledge creation generated by simultaneous co-operation and competition.

However, the historical development of buyer–supplier relations may also be analyzed better within the context of the national culture. Hofstede's (1984) measures showed Turkish culture to be relatively high in power distance and collectivism. Schwartz's (1994) measures similarly reflected a culture that emphasized tight links with the in-group and hierarchical roles for maintaining societal order. Turkish organizations are distinguished by centralized decision-making, highly personalized, strong leadership, and limited delegation (Ronen 1986). Turkish managers, likewise, are known for their autocratic and paternalistic styles (Pasa et al. 2001).

Moreover, sometimes, buyers and suppliers may not sufficiently communicate with each other about other significant sourcing and production variables as design, faster time to market, quality, and innovation, which are all crucial to supply-based competitiveness. We can say that the high degree of state involvement in business activity, be it in the form of subsidized credits, input supply or output demand, has been detrimental to the Turkish business environment.

Table 15.3 Benefits of a supply network: Turkish case

Critical element	Source of advantage	Characteristics
Product design and innovation	Regional cooperation and collaborations between supplier and buyers is encouraged in order to sustain competitive advantages and innovative aspects; if there are fewer suppliers, they must have complementary capabilities for buyers.	Design management is essential for enterprises. Synchronously, enterprises must follow efficient marketing and branding strategy through GVCs.
Manufacturing scale	Higher volumes of demand from global customers in a GVC perspective may enable manufacturers (and suppliers) to achieve the optimal production scale.	Enterprises must encourage to use commercial capital in order to be a part of transnational companies' supply chain.
Manufacturing factor costs	It is convenient to exert strategies to develop some certain competitive advantages from industrial locations (for example, low-cost producing countries in a GVC).	Enterprises must designate core competencies and enter R&D and global productions networks to reduce costs.
Design for manufacturability	Earlier supplier selection increases the level of strategic knowledge transfer in order to create designs that are faster, easier, and less costly to manufacture.	Non-durable consumer goods play an essential role in transfering knowledge among GVC.
Lean flow	Cooperation among a supplier and a buyer may simply reduce production and logistics costs.	Local enterprises must network among developed countries.
Transaction costs	Fewer transactions with fewer suppliers and more common terms of contracts significantly reduce cost.	Transnational corporations seek for low cost producers and suppliers.

Source: authors

Up till now, we have tried to argue how the emerging form of production organization does exist within the Turkish industrial districts in terms of relations with buyers, suppliers and other local and international producers. Moreover, we examined how these ties are encouraged in the process of upgrading of skills, technologies and products. In order to address these determinations, Table 15.3 reviews the benefits of a supply network for the Turkish manufacturing firms under the assumptions discussed above.

To sum up, we have argued that the presence of raw material suppliers and input manufacturers within the regional networks was cited to be a key locational advantage by Turkish manufactures. Moreover, most of the large firms in these districts have also reported to be relying upon local and global input suppliers. Among some of them, large firms are *vertically* integrating the production; in contrast, we may claim that most of the SMEs in the regional networks (or clusters) remained reliant on the local supplier and subcontracting networks. Furthermore, in contrast to SMEs, subcontracting and the local presence of input suppliers is examined in the case of specific literature on lower costs, generate externalities as playing an important role in the process of diffusing knowledge throughout the production network. On the other hand, buyers, particularly those representing international retailers, have an important role in the Turkish industrial districts. As they have acquired substantial technical expertise in the every related industry, this provides them the flexibility to be experienced marketing intermediaries (even to become a source for technical know-how in the production network).

Moreover, we have already denoted that some firms are bounded to (and embedded into) several GVCs providing further opportunities for linking other local enterprises that are in any kind of economic relation with them. Such firms are termed to be focal firms acting as the leading firms in the local innovation network, generating new knowledge and technologies, spinning out innovative companies, attracting researchers, investments and research facilities, enhancing other firms' R&D activities, stimulating demand for new knowledge and creating and capturing externalities (Agrawal and Cockburn 2002; Boari and Lipparini 1999; Lazerson and Lorenzoni 1999; Saxenian 1991). Parallel to the new stages of learning and innovation to achieve the goal of industrial development, finally, we may well advance the hypothesis that the presence of focal firms in production network substantially increases spillovers at the local level, by creating technologically-advanced new knowledge and favoring the absorption and dissemination of external knowledge into the network parallel to the theory that GVC offers spontaneous technological and economic structures to link local firms to global networks.

15.4 International Production Networks: Types of Knowledge and Knowledge Spillovers

Typically, the knowledge base of traditional industries is highly dependent upon local and tacit forms of knowledge, whereas the knowledge base of firms in high-technology sectors is more codified allowing firms to establish networks to access distant knowledge sources (Vale and Caldeira 2008). However, in the most of the prominent work done by economics researchers, the divide between local/tacit knowledge and non-local/codified knowledge has been criticized (Gertler 2003). There are still reports of poor transactions at the inter-company level within networks, as well as examples of companies that do not rely only on local sources to innovate; rather they will often consistently establish distant networks in order to access new knowledge and combine it with local assets.

As a well known economic fact, firms dispose of capabilities to store and develop knowledge through their rules and routines as well as through specific documentation procedures, as Nelson and Winter (1982) have shown. In recent approaches to the theory of the firm, enterprises have been considered not only as repositories of knowledge, but also as processors of knowledge (Amin and Cohendet 2000).

In the development of firms and regions, the significance of tacit knowledge and codified knowledge has been extensively discussed. Occasionally, a simplified dualism is assumed where tacit knowledge is considered to be in-replicable, providing regions and firms with a continuous advantage of innovation and capability building, while codified knowledge is considered to be clearly available because of its standardization, replicability and codification properties. Consequently, this kind of knowledge is also assumed to create strong regional and global competitiveness powers. Meanwhile, however, more complex typologies of knowledge transfer and organizational learning along the dimensions of tacit versus codified (and individual versus collective) knowledge were recently developed (e.g., Amin and Cohendet 1999; Gertler 2003).

Furthermore, as one of the most important explanations of why innovative activity is geographically concentrated is that knowledge is a crucial element of innovation (Simmie 2002). Here, knowledge, particularly tacit knowledge, spillovers from individual firms and institutions to others in the same place. We may also argue that the successful knowledge transfer happens along in a distance. It is therefore argued that spatial concentrations of knowledge-rich firms and institutions benefit from knowledge spillovers.

We must also denote that the success of organizational learning depends on the firms' absorptive capacity, which itself is determined by the firm's prior related knowledge (see Kim 1998). Here, the definition of knowledge refers to the recipient firms' ability to recognize the value of new knowledge or information, assimilate it, and apply it to commercial ends (Daghfous 2004). Such an action was theoretically labeled as "absorptive capacity" by Cohen and Levinthal (1990). In this regard, recent studies showed us that the knowledge created within firms in an industrial district can be used by other economic agents, because pieces of that knowledge can be codified and transferred among firms; thus generating positive externalities and fostering innovative activities.[2] Extending this body of research with a greater attention to the specificities of knowledge flows and their impact at the firm level (Malerba 2005), knowledge spillovers have been defined as public good bounded in space (Breschi and Lissoni 2001).

According to this approach, most of the knowledge flowing is mainly "tacit", context specific and difficult to codify, and this is particularly true for innovative ideas. As a consequence, it can be primarily transmitted through personal contacts and direct inter-firm relationships. Following the "Marshallian" concept of industrial districts, it is also argued that such knowledge flows better among organizations located in the same area (Krugman 1991). Therefore, networked firms have more innovative advantages and opportunities than a scattered location (Breschi and Lissoni 2001; Saxenian 1994), and firms located in regions

---

[2] See Griliches (1979) for the basic theory.

characterized by knowledge-agglomeration processes have greater opportunity to access this knowledge than their distantly located competitors.

Consequently, while there were technical limitations that prevented the conventional approaches from unveiling the underlying complex inter-firm relationships and knowledge spillovers in detail, first, social network analysis offered a methodological breakthrough to overcome such limitations (see Nakato 2004).

As a preliminary draft for understanding the business structure of Turkey, we may depict that the country achieved a lowered ranking of 58th in the business sophistication pillar of the Global Competitiveness Index (GCI), particularly for the quality and quantity of networks and supporting industries, below the EU average, and below the states of developing countries like Estonia, the Czech Republic, and Slovenia (World Economic Forum 2012). According to us, this scheme strongly suggests that while Turkey does have a large agricultural sector with rather low productivity, both in relation to the agricultural sector of other recent EU entrants and in relation to other sectors in the Turkish economy; having sophisticated industrial and service sectors; we may not argue whether enterprises are operating at high levels of efficiency, adopting advanced technologies and efficient production processes, nor exploiting economies of scale with respect to their competitors elsewhere in Europe, e.g. compared to the new members in central and Eastern Europe. In this respect, the larger the scale of exploitation is, in the developing countries case, we can depict that the social structure among agents (individuals and/or firms) must create the pre-conditions for innovation by building up relational networks in the GVCs.

In this manner, we may also argue that Turkish SMEs' business activities are strongly influenced by the social structure. Accordingly, the networks of relations among them have certainly developed in the entangled chains of manufacturing processes in an organized and complex web of geographically bound, subcontracting business networks. As when a different variety of firms from different sectors were embedded in the Turkish regional manufacturing systems, firms develop new inter-organizational relationships for the spillover of knowledge and technology in the industrial district they facilitate. Some of the underlying structural and relational patterns may be sorted as Turkish manufacturing firms are embedded in the regional business networks; trust and informal relations are so important in the context of business relations.

From the current research, we can clearly define a new range of options to make international comparisons. In the Turkish case, we may depict that there is no common and unidirectional development pattern which has been followed by the new different competitive challenges posed by the globalization of markets and technology. As denoted in the previous part, by the variety of visions on the notion of industrial districts in the literature, we may also depict the existence of some enterprise clusters and agglomerations that have been recorded worldwide.

Continuously, an examination of the broad characteristics of the Turkish business environment shows that SMEs account for more than 90 % of Turkish firms, but larger firms' contribution to value-added and exports are much greater (Taymaz 1997). Big corporations are a relatively new phenomenon in Turkey: of the

405 TUSIAD member companies, only 22 were established before 1950 (Buğra 1994). The 1950s were an important decade for many of the largest Turkish companies, reflecting the government's shift to more liberal policies. Many of today's leading Turkish construction firms, for example, were either established or made an important turn in their business during that decade (Öz 1999).

Moreover, family-dominated management of firms of all sizes is common for Turkish industry as there is a lack of confidence in salaried managerial personnel Educating young members of the family in top universities, integrating a professional manager into the family via marriage, and strong relationships established over the years between family members and professional managers, making the latter 'part of the family', appear to be usual ways of achieving a delicate balance between professionalization and family control (Buğra 1994).

According to Buğra (1994), all Turkish business tycoons have certain characteristics in common, including family support in commercial activities at the start of their life-cycle, the arbitrary choice of their initial area of activity, heavy engagement in unrelated diversification as the business grows, and good business relations especially in state circles. Moreover, we might denote that the high degree of state involvement in business activity (in the form of subsidized credits input supply or output demand) has been detrimental to the Turkish business environment. Furthermore, given the key role of government in the economy, we may argue that good connections in governmental approaches have contributed significantly to business successes. The slow bureaucracy and unexpected changes in key policies, on the other hand, have caused problems for the Turkish business community.

One another aspect is that public funding from governmental bodies like TÜBİTAK (Scientific and Technological Research Council of Turkey) and DPT (State Planning Organization) is to be effectively translated into marketable products and services. The role of businesses is crucial to strengthen the technological and innovation performance of enterprises that will eventually tend to support knowledge transfer from other networks of organizations. In theoretical conditions, knowledge transfer requires the right economic environment to support and stimulate business to link with suppliers, customers and the research base. These linkages will primarily be created and financed by industry. But, we would like to mention that there is a key role for the Turkish national government to help managing business markets in particular activities or regions, and investing strategically in new strands of science and technology. In this regard, the private sector must also interact with university research. As equally, universities and the public sector must assess the realistic opportunities for the commercial exploitation of their research, and an understanding of the priorities and needs of the private sector.

15.5 Summary

In this article, we have presented some clues for the developing countries based on GVC and GCI index in such a reasoning that entering GVCs may not provide an automatic move up the capability ladder. The process must start with a fast track recording in regional networks to acquire new production capabilities. In the Turkish case, we may see relative explanations for some enterprises to have their capabilities downgraded as a result of their integration in global value chains. So, it makes sense for latecomers to use all the resources they can acquire first from regional networks and on the following, from the developed countries, in return for providing such services as low-cost manufacturing. But, one must not forget that the services tradeoff can be exploited to the advantage of the developing countries only if there is a strategic choice to use the links to gain knowledge to learn.

Moreover, innovation activities within GVCs may move along two dimensions of leverage strategies: services expansion in regional networks and acquisition of technological capabilities from developed countries.

Hence, in this generic scheme, we can say that Turkey is favored by its large internal markets, but also shows the benefits of the recent microeconomic reforms promoting regional networks and global competition, simultaneously. As also demonstrated by the variety of product specializations of SMEs in Turkey, we may also argue that the degree of complexity of organizational and network systems and the scope—variety of inter-firm Turkish organizations are continuously expanding, in relation to the globalization of technology and the increasing inter-nationalization and localization of economic activities; but, not at the desired levels of inclusion to GVCs compared with the Global Competitiveness Index business sophistication statistics.

Finally, we conclude that integrating an enterprise or local cluster into a GVC is an important step, nevertheless, the SMEs or clusters do not have to see their horizons limited. Enterprises must always seek ways of spreading their involvement across two or more GVCs, as they have to expand their opportunities and capabilities, simultaneously. Only by this way, enterprises may leverage skills, enhance capabilities and reduce the risk of being tied to a single.

References

Agrawal A, Cockburn IM (2002) University research, industrial R&D and the anchor tenant hypothesis. In: NBER Working Paper 9212. National Bureau of Economic Research, Cambridge

Akgüngor S (2003) Exploring regional specializations in Turkey's manufacturing industry. In: Paper prepared for presentation at the Regional Studies Association International Conference on Gateway 7: Regional Competitiveness. Pisa, Italy, April 12–15, 2003

Altenburg T (2006) Governance patterns in value chains and their development impact. Eur J Dev Res 18:498–521

Amin A, Cohendet P (1999) Learning and adaptation in decentralised business networks. Environ Plan D 17:87–114

Amin A, Cohendet P (2000) Organizational learning and governance through embedded practices. J Manag Gov 4:93–116

Armatli-Köroğlu B, Beyhan B (2003) The changing role of SMEs in the regional growth process: the case of Denizli. In: Fingleton B, Eraydin A, Paci R (eds) Regional economic growth. SMEs and the Wider Europe. Ashgate, London, pp 229–245

Asanuma B (1989) Manufacturer-supplier relationships in Japan and the concept of relation-specific skill. J Japanese Int Econ 3:1–30

Bathelt H (2003) Geographies of production: growth regimes in spatial perspectives on innovation, institutions and social systems. Prog Human Geogr 27(6):789–804

Boari C, Lipparini A (1999) Networks within industrial districts: organizing knowledge creation and transfer by means of moderate hierarchies. J Manag Govern 3:339–360

Breschi S, Lissoni F (2001) Knowledge spillovers and local innovation systems: a critical survey. LIUC Papers in Economics 84

Buğra A (1994) State and business in modern Turkey: a comparative study. State University of New York Press, New York

Cohen MD, Levinthal DA (1990) Absorptive capacity: a new perspective on learning and innovation. Admin Sci Quart 35:128–152

Daghfous A (2004) An empirical investigation of the roles of prior knowledge and learning activities in technology transfer. Technovation 24:939–953

Dede OM (1999) Spatial structure of technology based production in Turkey: the case of professional electronics industry in Ankara. PhD dissertation, METU, Ankara

Doloreux D, Parto S (2005) Regional innovation systems: current discourse and unresolved issues. Technol Soc 27:133–153

Eraydin A (1998) The role of regulation mechanisms and public policies at the emergence of the new industrial districts. In: Paper presented at the Symposium on New Nodes of Growth in Turkey: Gaziantep and Denizli, Ankara

Eraydin A, Armatlı-Köroğlu B (2005) Innovation, networking and the new industrial clusters: the characteristics of networks and local innovation capabilities in the Turkish industrial clusters. Entrep Reg Dev 17:237–266

Erdil E, Göksidan HT (2006) Inter-organizational relations in an organized industrial district: Ostim case. Unpublished paper from MSc Thesis

Erendil A (1998) Using critical realist approach in geographical research: an attempt to analyze the transforming nature of production and reproduction in Denizli. Unpublished PhD thesis, Department of City and Regional Planning, Middle East Technical University, Ankara

Ersoy M (1993) Yeni Liberal Politikalar ve Kentsel Sanayi. METU Faculty of Architecture, Ankara

Gereffi G (1994) The organization of buyer-driven global commodity chains: how US retailers shape overseas production networks. In: Gereffi G, Korzeniewicz M (eds) Commodity chains and global capitalism. Praeger, London

Gereffi G (1999) International trade and industrial upgrading in the apparel commodity chain. J Int Econ 48:37–70

Gereffi G, Kaplinsky R (2001) The value of value chains. IDS Bull 32(3)

Gereffi G, Korzeniewicz M (eds) (1994) Commodity chains and global capitalism. Praeger, Westport

Gertler MS (2003) Tacit knowledge and the economic geography of context, or the undefinable tacitness of being (there). J Econ Geogr 3:75–99

Giuliani E (2005) When the micro shapes the meso: learning and innovation in wine clusters. PhD Thesis. SPRU, University of Sussex, Brighton

Griliches Z (1979) Issues in assessing the contribution of R&D to productivity and growth. J Econ 10:92–116

Henderson J, Dicken P, Hess M, Coe N, Yeung HYC (2002) Global production networks and the analysis of economic development. Rev Int Politic Econ 9(3):436–464

Hofstede G (1984) Culture's consequences: international differences in work-related attitudes. Sage, Beverly Hills

Humphrey J, Schmitz H (2002a) How does insertion in global value chains affect upgrading industrial clusters? Region Stud 36:1017–1027

Humphrey J, Schmitz H (2002b) Developing country firms in the world economy: governance and upgrading in global value chains. INEF Report No. 61, University of Duisburg, Duisburg

Humphrey J, Schmitz H (2004) Governance in global value chains. Institudo Nacional de Technologia Industrial (INTI), Spain

Kaplinsky R (2000) Globalisation and unequalisation: what can be learned from value chain analysis? J Dev Stud 37:117–146

Kim L (1998) Crisis constructon and organizational learning: capability builing in catching up at Hyundai motor. Org Sci 9:506–521

Krugman P (1991) Geography and trade. MIT Press, Cambridge, MA

Kuruüzüm O (1998) Bilimsel Bilginin Ticarileştirilmesinde Yeni Ufuklar: Teknoloji Geliştirme Bölgeleri. Akdeniz Üniversitesi Yayınları

Lall S (2001) Competitiveness indices and developing countries: an economic evaluation of the global competitiveness report. World Dev 29(9):1501–1525

Lazerson M, Lorenzoni G (1999) The firms that feed industrial districts: a return to the Italian source. Ind Corporate Change 82(2):235–266

Malecki E, Oinas P (eds) (2000) Making connections: technological learning and regional economic change. Ashgate, Aldershot

Malerba F (2005) Sectoral systems: how and why innovation differs across sectors. In: Fagerberg J, Mowery D, Nelson R (eds) The Oxford handbook of innovation. Oxford University Press, Oxford

Morrison A, Pietrobelli C, Rabellotti R (2008) Global value chains and technological capabilities: a framework to study learning and innovation in developing countries. Oxford Dev Stud 36 (1):39–58

Nakato T (2004) Bridging roles of SMEs in a large scale industrial district: a structural approach. University of Michigan Flint School of Management Working Paper Series

Nelson R, Winter S (1982) An evolutionary theory of economic change. Harvard University Press, Cambridge

Öz Ö (1999) The competitive advantage of nations: the case of Turkey. Ashgate Publishing, Farnham

Öz Ö (2004) Clusters and competitive advantage: the Turkish experience. Palgrave Mac-Millian, New York

Özcan GB (1995) Small firms and local economic development: entrepreneurship in Turkey. Avebury, Aledershot

Pasa S, Kabasakal H, Bodur M (2001) Society, organizations, and leadership in Turkey. Appl Psychol Int Rev 50(4):59–89

Peres W, Stumpo G (2000) Small and medium-sized manufacturing enterprises in Latin America and the Caribbean under the new economic model. World Dev 28(9):1643–1655

Pietrobelli C (1998) Industry, competitiveness and technological capabilities in Chile, A new tiger from Latin America? MacMillan and St. Martin's, London and New York

Pietrobelli C, Rabellotti R (2007) Upgrading to compete. Global value chains, clusters and SMEs in Latin America. Harvard University Press, Cambridge

Porter M (1990) The competitive advantage of nations. Macmillan, London and Basingstoke

Reyhan N (1990) The spatial implications of restructuring of production organization of Bursa textile industry. MSc dissertation, METU, Ankara

Ronen S (1986) Comparative and multinational management. Wiley, New York

Saliola F, Zanfei A (2009) Multinational firms, global value chains and the organization of knowledge transfer. Res Pol 38:369–338

Saraçoğlu Y (1993) Local production networks: an opportunity for development. MSc dissertation, METU, Ankara

Saxenian A (1991) The origins and dynamics of product networks in Silicon Valley. Res Pol
    20:423–437
Saxenian A (1994) Regional advantage: culture and competition in Silicon Valley and Route 128.
    Harvard University Press, Cambridge
Schwartz SH (1994) Beyond individualism/collectivism: new cultural dimensions of values. In:
    Kim U et al (eds) Individualism and collectivism: theory, methods, and applications. Sage,
    Thousand Oaks, CA, pp 85–119
Simmie JM (2002) Trading places: competitive cities in the global economy. Eur Plan Stud 10
    (2):201–214
Taymaz E (1997) The dynamics of firms in a micro-to-macro model with training, learning, and
    innovation. J Evolut Econ 7:435–457
Tekeli İ (1994) Ankara'da tarih icinde sanayinin gelişimi ve mekansal farklılaşma. Yapi Kredi
    Yayinlari, Ankara, pp 171–200
Ulusoy G (2003) An assessment of supply chain and innovation management practices in the
    manufacturing industries in Turkey. Int J Prod Econ 86:251–270
Vale M, Caldeira J (2008) Fashion and the governance of knowledge in a traditional industry: the
    case of the footwear sectoral innovation system in the northern region of Portugal. Econ Innov
    New Technol 17(1–2):61–78
World Economic Forum (2011) The global competitiveness report 2011–2012. Geneva

Part IV

Conclusions

Seizing Opportunities for National STI Development

16

Leonid Gokhberg and Dirk Meissner

Science, technology and innovation (STI) is broadly assumed to be among the main drivers of change in contemporary economies and societies. Accordingly, it is recognized and accepted that STI contributes to addressing national challenges and problems. But this raises an important issue: does framing matters in terms of challenges and problems providing an effective way of mobilizing resources? This sort of framing may actually be productive in some circumstances, for some stakeholders, but less so in other contexts and with other agents of change.

In particular, there is a certain concern with how scientists and engineers perceive research. Typically, they are ambitious in their efforts to solve a problem. Thus, initially they describe and decompose the problem to uncover all its possible facets and fully understand it. An activity to solve the problem follows in the tradition of scientific work. This approach is targeted at directing efforts to each feature of the problem and finding a solution for this. Each solution to the sub-components of the wider problem is in most cases treated independently, without being incorporated into an overarching consistent system. The reason is that problems are now typically larger in scale and more complex compared to just a few decades ago. This means that many scientific working groups must cooperate to solve challenges and issues, even though these teams usually compete among themselves (Gokhberg and Sokolov 2013; OECD 2011; Schibany and Reiner 2014; Meissner 2015).

This practice of scientific work is undoubtedly productive for understanding problems and developing new knowledge. However, the results are very sophisticated and specialized which means that their ability to be integrated into broader systems is limited. It is broader systems indeed that are in demand for solving the broader challenges. In this respect, there is a clear need to shift in the ways how we

L. Gokhberg (✉) • D. Meissner
Institute for Statistical Studies and Economics of Knowledge, National Research University
Higher School of Economics, 20 Myasnitskaya Street, 101000 Moscow, Russia
e-mail: lgokhberg@hse.ru; dmeissner@hse.ru

© Springer International Publishing Switzerland 2016
L. Gokhberg et al. (eds.), *Deploying Foresight for Policy and Strategy Makers*,
Science, Technology and Innovation Studies, DOI 10.1007/978-3-319-25628-3_16

perceive and solve problems. 'Thinking in Opportunities' instead of 'Thinking in Problems' is not only a play with words but has practical implications of a strategic nature.

'Thinking in Opportunities' implies that an issue is treated in a way which specifies the requirements for a potential solution. These requirements are matched against existing solutions which might be appropriate to solving the problem. The next step is analyzing the gaps which arise from matching requirements and potential solutions, and further decomposing these gaps into smaller issues and related solution requirements until it is possible to clearly formulate research projects and plans to solve the overarching problem. A special feature of this 'solution-driven approach' (also called the 'opportunity-driven approach') is that from a very early stage the interfaces between different features and components of the overall system are considered. However, any challenge or problem is unique and consequently the solutions are varied. Hence the 'Thinking in Opportunities' approach that we have outlined here may be applied broadly but it needs to be elaborated in much more detail, particularly for the purposes of designing an STI policy mix or corporate strategy.

Opportunity-driven thinking is a widespread motivation for company innovation activities. Therefore, a problem **and** opportunity analysis is a more widely used instrument in innovation management and business development than in scientific communities. Another challenge however is that solutions—and hence innovations—vary not only in shape (product, process, service, business model etc.) but also in the underlying competencies required for a given application or technology field. While the same sources of innovation (whether that is the science and research base or commercial entities) pertain, their relative weights are changing. Opportunity thinking requires a stronger orientation to applicable solutions rather than the approach which aims to more fully explore a problem ('problem thinking'). Therefore the requirements of users for a solution (e.g. the companies who possess the competencies to identify and address certain problems) gain more weight as sources for innovation, compared to the science and research base (Brown 2003; Geroski 2000; Reed et al. 2009).

Consequently, the share of pull innovation increases while that of push innovation decreases. Solutions to meet challenges or solve problems are however more about push innovation by nature than the pull one because a user of the innovation is not necessarily known to anyone at the current stage. Therefore, a mixed STI policy approach is reasonable as it would reconsider the balance between different innovation sources in light of the specific characteristics of the challenges. The interfaces between different individual solutions which need to be integrated into complex systems require more careful elaboration in the early phases of STI policy design in order to develop a smooth and seamless STI policy mix (Gokhberg and Meissner 2013; Meissner 2014).

'Thinking in Opportunities' in fact is very different from the 'Thinking in Problems' approach because it predominantly looks forward and is concerned with creating innovations. In contrast, the problem-solving model mostly addresses existing systems and thus often looks backwards (i.e. it addresses problems of the past).

The opportunities-orientated approach fits much better with the nature and contents of foresight studies, in particular in the field of STI. Taking account of multiple factors that impact potential catching up opportunities and a broad spectrum of relevant research problems, this approach—at the national level—creates a backdrop for launching large-scale, complex STI projects that can be funded i.e. through public-private partnership schemes (Gokhberg and Meissner 2013).

Nevertheless, this new paradigm for STI policy applies to all countries irrespective of the developmental status of their National Innovation System (NIS) (Meissner et al. 2013). We are used to hearing that to become or remain globally competitive, countries need to boost national innovativeness. However, this simple formulation is liable to be counterproductive, if we do not disentangle what is meant by 'competition'. Policy measures aimed at increasing national competitiveness need to consider the different dimensions of competition, among which the following are central to STI:

- **Global industrial competition** is the traditional competition faced by companies for the best solution to user requirements at the best price; for some authors the only form of competition is that between companies, but in reality the next two categories of 'competition' are often also brought into play.
- **Global science and research base competition** is frequently understood as the international competition for the outputs from science and research. Recently, international *competition for talent* has become more intensive as an increasing numbers of countries have begun to promote their national STI systems to achieve, first, excellence in science and, second, excellence in innovation. This has led to an intensified *competition for favorable research and innovation environments* to make the NIS attractive for talent and investment both globally and domestically. It mainly targets the relevant framework related policy instruments, including labor and migration laws and tax incentives, as well as major STI support mechanisms such as funding and remuneration schemes, etc.
- **Global competition of countries** leads to efforts to design the best possible framework conditions for entrepreneurship and innovation which are often generous, in particular towards companies' investment in national STI related activities.

To increase the efficiency of national economies and STI systems, most countries have been developing national innovation strategies. These strategies frequently involve STI priority setting, smart specialization, public-private partnerships, mechanisms promoting industry-science linkages, cluster policies and technology platforms, tax credits and other subsidies, as well as earmarked measures to attract and keep talent. Setting priorities for STI is one of the most burning issues for national governments. Frequently, foresight is used to identify and set up related priorities. However, foresight often has a rather narrow focus on STI, neglecting societal and environmental developments (Georghiou 2013; Simachev et al. 2014; Kasimov et al. 2015).

Although many countries have—at least partially—developed and implemented national STI strategies, these strategies are increasingly challenged. In any strategy, there are always winners and losers in a NIS, and political establishments play an important role. STI strategies in particular have been often perceived as suitable policy initiatives by policy makers who have very personal agendas. Moreover, the political establishment in most countries is characterized by the competition for influence between national entities (ministries, agencies, councils, etc.). For example, national and regional institutions might not follow a shared ambition and vision which ultimately limits the possibilities for a coherent and consistent strategy. In addition to challenges in developing a strategy, implementing national STI strategies is often difficult. Although top-level policy initiatives create a certain momentum among STI actors for some time, the actual implementation of policy actions (either designed in the national strategy or derived from the strategy's overarching directions) is usually left to subordinated agencies. In this regard, experience shows that the more agencies are involved in implementing STI, the less stringent and sustainable the implementation of strategic measures is. The reason for this is the different perceptions and interpretations of strategies by the implementing agencies, which is at least partially due to their respective roles and duties.

National STI strategies need to take into account the potential future developments of society, industry, science and policy under a variety of possible scenarios. The latter are important because countries' stated priorities often alter when the leadership and/or socio-economic conditions change. An important determinant for the successful implementation of STI strategies is the institutional organization of the NIS. In principle, the organizations making up the NIS should follow the main strategic intentions. For example, the strategies require complementary institutional adjustments. Moreover, the institutions' structure alone does not guarantee a desirable impact and success; instead, the success requires communications and education of the people involved as well as near time operational measures to maintain the initial momentum (Gokhberg 2013; Meissner 2014).

Ultimately, a STI policy mix which follows the opportunity thinking approach and tackles challenges needs to look beyond its traditional elements. *Competition and trade policies* need to work in tandem to discourage rent-seeking behaviour and help economic actors in accessing global markets and communities. *Finance and investment policies* should focus on supporting financial institutions that are able to properly value innovation-related investment. In addition, this includes supporting the efficient management of some of the risks inherent to smart specialization and resulting innovation within specialized clusters. *Education and training policy,* together with *labour market policy*, should help secure the quantity, quality and efficient allocation of human resources, while *research policy* needs to be targeted at developing and mobilising mutually reinforcing research capabilities in the public and private sectors. *Industrial and regional policy instruments* need to develop and maintain an appropriate infrastructure and other support mechanisms to realise the innovation potential of specific sectors. *Social and health policy* should consider innovation as a means—as well as a result—to improve the quality of life. *Environmental policy* views pro-innovation regulations and incentives as important tools to encourage value-creating responses to the sustainability challenge. Finally, *legal policy* exists to enforce the rule of law, protecting intellectual property rights and broader innovation activities that are already inherently risky against additional, unbearable uncertainties (European Communities 2002).

These policy fields require increased and improved interaction to ensure sustainable functioning of the NIS. It is common for these policy measures to be developed and implemented by dedicated governmental agencies. In most cases, the different policy measures do not fit together well. They are designed towards piecemeal objectives of various public agencies and usually made at different times, which can also be counterproductive to a coherent STI policy mix. A coherent STI policy mix that 'supports opportunities' demonstrates a number of characteristics, namely:

- It adheres to an anticipatory model to address the most essential, systemic failures and leads STI advancement and utilization to contribute to economic growth, inclusiveness, and green/sustainable development including a strong commitment to support STI in the sectors that are important to enhance competitive advantages at national and regional levels, to stimulate youth creativity, and are prepared to pro-innovation attitudes.
- It recognizes that national STI development is a long-term undertaking which also requires significant improvements in national STI strategies including forward looking priority setting and the governance system.
- It emphasizes the regular use of foresight and allied strategic intelligence tools which take account of industrial, technological, and scientific developments, market and application field-specific trends, as well as significant STI and related policy process developments, thereby stressing the implementation of foresight in national STI strategies and the succeeding development and implementation of STI policy measures.
- It elaborates national STI strategies with special emphasis on partnerships which are required to establish the balance between basic research, applied research, and commercial interests to assure a reasonable pipeline for innovation in the future. In this respect, competing values of potential partners are involved which have to be accounted for and the respective incentives set for partners to disclose their related strategic intentions and limit (or even avoid) the rent-seeking interests of individual partners.
- It broadens the traditional linkages between NIS actors, e.g. the collaboration and partnership paradigm, by considering the existence and the power of global STI networks and the global value chains of industrial sectors, hence involving horizontal and vertical linkages between the actors.
- It reflects on the role of public authorities (such as regional or national governments and affiliated bodies) as coordinators and also players in enabling the process of entrepreneurial discovery and in designing public or semi-public institutions, setting framework policies and standards affecting technological attributes, user demand and other market factors in designing industrial policies;

- It creates awareness among researchers in universities and public research institutes about research management and strategic research orientation, especially an awareness of the relationship between results and applications, and commitments to open communications with the society and partners, which determines the quality and design of information, communication and decision-making processes.

The implementation of policy initiatives requires effective governance, which in turn often leads to emerging skills and training needs for employees of regional and national governments. Furthermore, it is essential to have a close interaction regarding STI policies and implementation measures across the different levels in a country to assure coherence between innovation strategies at different levels, 'translate' regional choices into terms used in the national strategy, and reach the targeted STI community in a country to achieve the intended impacts.

Ultimately, the design and implementation of a consistent and coherent STI policy mix which clearly addresses the features of 'Thinking in Opportunities' is crucial for countries to generate momentum and take advantage of the full potential of STI.

The editors wish to express their gratitude to all the contributors to this book. The different book chapters together provide a wide-ranging overview, and contribute to in-depth discussions of many different facets of foresight and STI policy and company strategy. Such overview and discussion help us in the effort to shift to a more positive opportunities-driven perspective.

Acknowledgements The chapter was prepared within the framework of the Basic Research Programme at the National Research University Higher School of Economics (HSE) and supported by a subsidy granted to the HSE by the Government of the Russian Federation for the implementation of the Global Competitiveness Programme.

References

Brown MM (2003) Technology diffusion and the "knowledge barrier": the dilemma of stakeholder participation. Public Perform Manag Rev 26(4):345–359

European Communities (2002) Innovation tomorrow—Innovation policy and the regulatory framework: making innovation an integral part of the broader structural agenda. Luxembourg, Office for Official Publications of the European Communities. http://ftp.cordis.europa.eu/pub/innovation-policy/studies/studies_innovation_tomorow.pdf, last accessed 27.07.2015

Georghiou L (2013) Challenges for science and innovation policy. In: Meissner D, Gokhberg L, Sokolov A (eds) Science, technology and innovation policy for the future—potentials and limits of foresight studies. Springer, Heidelberg, New York, Dordrecht, London, pp. 233–246

Geroski PA (2000) Models of technology diffusion. Res Pol 29:603–625

Gokhberg L (2013) Indicators for science, technology and innovation on the crossroad to foresight. In: Meissner D, Gokhberg L, Sokolov A (eds) Science, technology and innovation policy for the future—potentials and limits of foresight studies. Springer, Heidelberg, New York, Dordrecht, London, pp 257–288

Gokhberg L, Meissner D (2013) Innovation: Superpowered invention. Nature 501:313–314. doi:10.1038/501313a

Gokhberg L, Sokolov A (2013) Targeting STI policy interventions—future challenges for fore-sight studies. In: Meissner D, Gokhberg L, Sokolov A (eds) Science, technology and innovation policy for the future—potentials and limits of foresight studies. Springer, Heidelberg, New York, Dordrecht, London, pp 289–292

Kasimov N, Alekseeva N, Chulok A, Sokolov A (2015) The future of the natural resources sector in Russia. Int J Soc Ecol Sustain Dev 6(3):80–103

Meissner D (2014) Approaches for developing national STI strategies. STI Pol Rev 5(1):34–56

Meissner D (2015) Developing 'Green Thinking' towards sustainability. Int J Soc Ecol Sustain Dev 6(3):iv–vii

Meissner D, Roud V, Cervantes M (2013) Innovation policy or policy for innovation?—in search of the optimal solution for policy approach and organisation. In: Meissner D, Gokhberg L, Sokolov A (eds) Science, technology and innovation policy for the future—potentials and limits of foresight studies. Springer, Heidelberg, New York, Dordrecht, London, pp 247–255

OECD (2011) Skills for innovation and research. OECD, Paris

Reed MS, Graves A, Dandy N, Posthumus H, Hubacek K, Morris J, Prell C, Quinn CJ, Stringe LC (2009) Who's in and why? A typology of stakeholder analysis methods for natural resource management. J Environ Manag 90:1933–1949

Schibany A, Reiner C (2014) Can basic research provide a way out of economic stagnation? Foresight-Russia 8(4):54–63

Simachev Y, Kuzyk M, Kuznetsov B, Pogrebnyak E (2014) Russia heading towards a new technology—industrial policy: exciting prospects and fatal traps. Foresight-Russia 8(4):6–23 (in Russian)